# Creo Parametric 6.0 Black Book

By
**Gaurav Verma**
**Matt Weber**
**(CADCAMCAE Works)**

Edited by
**Kristen**

ISBN # 978-1-988722-56-6

NOTICE TO THE READER

Publisher does not warrant or guarantee any of the products described in the text or perform any independent analysis in connection with any of the product information contained in the text. Publisher does not assume, and expressly disclaims, any obligation to obtain and include information other than that provided to it by the manufacturer.

The reader is expressly warned to consider and adopt all safety precautions that might be indicated by the activities herein and to avoid all potential hazards. By following the instructions contained herein, the reader willingly assumes all risks in connection with such instructions.

The Publisher makes no representation or warranties of any kind, including but not limited to, the warranties of fitness for a particular purpose or merchantability, nor are any such representations implied with respect to the material set forth herein, and the publisher takes no responsibility with respect to such material. The publisher shall not be liable for any special, consequential, or exemplary damages resulting, in whole or part, from the reader's use of, or reliance upon, this material.

## DEDICATION

To teachers, who make it possible to disseminate knowledge
to enlighten the young and curious minds
of our future generations

To students, who are the future of the world

## THANKS

To my friends and colleagues

To my family for their love and support

# Training and Consultant Services

At CADCAMCAE WORKS, we provides effective and affordable one to one online training on various software packages in Computer Aided Design(CAD), Computer Aided Manufacturing(CAM), Computer Aided Engineering (CAE), Computer programming languages(C/C++, Java, .NET, Android, Javascript, HTML and so on). The training is delivered through remote access to your system and voice chat via Internet at any time, any place, and at any pace to individuals, groups, students of colleges/universities, and CAD/CAM/CAE training centers. The main features of this program are:

## Training as per your need

Highly experienced Engineers and Technician conduct the classes on the software applications used in the industries. The methodology adopted to teach the software is totally practical based, so that the learner can adapt to the design and development industries in almost no time. The efforts are to make the training process cost effective and time saving while you have the comfort of your time and place, thereby relieving you from the hassles of traveling to training centers or rearranging your time table.

## Software Packages on which we provide
## basic and advanced training are:

CAD/CAM/CAE: CATIA, Creo Parametric, Creo Direct, SolidWorks, Autodesk Inventor, Solid Edge, UG NX, AutoCAD, AutoCAD LT, EdgeCAM, MasterCAM, SolidCAM, DelCAM, BOBCAM, UG NX Manufacturing, UG Mold Wizard, UG Progressive Die, UG Die Design, SolidWorks Mold, Creo Manufacturing, Creo Expert Machinist, NX Nastran, Hypermesh, SolidWorks Simulation, Autodesk Simulation Mechanical, Creo Simulate, Gambit, ANSYS and many others.

Computer Programming Languages: C++, VB.NET, HTML, Android, Javascript and so on.

Game Designing: Unity.

Civil Engineering: AutoCAD MEP, Revit Structure, Revit Architecture, AutoCAD Map 3D and so on.

We also provide consultant services for Design and development on the above mentioned software packages

For more information you can mail us at:
cadcamcaeworks@gmail.com

# Table of Contents

## Chapter 3 : Advanced Sketching and Practical

## Chapter 4 : 3D Modeling Basics

## Chapter 5 : 3D Modeling Practical and Practice

## Chapter 6 : 3D Modeling Advanced

## Chapter 7 : 3D Modeling Advanced Practical and Practice

## Chapter 8 : 3D Modeling Advanced-II

## Chapter 11 : Sheetmetal Design

## Chapter 12 : Surface Design

## Chapter 13 : Drawing

## Chapter 14 : Model Based Definition (MBD) and 3D Printing

## Chapter 15 : Introduction to Sheetmetal Manufacturing

# Preface

Creo Parametric 6.0 is a parametric, feature-based solid modeling tool that not only unites the three-dimensional (3D) parametric features with two-dimensional (2D) tools, but also addresses every design-through-manufacturing process. The continuous enhancements in the software has made it a complete PLM software. The software is capable of performing analysis with an ease. Its compatibility with CAM software is remarkable.

To support education, PTC gives free educational version of the Creo Parametric to students with a license of 1 year. You can get your free version of software by visiting the following link:

http://www.ptc.com/communities/academic-program/products/free-software

The **Creo Parametric 6.0 Black Book** is the fourth edition of our series on Creo Parametric. With lots of additions and thorough review, we present a book to help professionals as well as learners in creating some of the most complex solid models. The book follows a step by step methodology. In this book, we have tried to give real-world examples with real challenges in designing. We have tried to reduce the gap between university use of Creo Parametric and industrial use of Creo Parametric. In this edition of book, we have included new enhancements of Creo Parametric 6.0 interface. We have included an introductory chapter on Sheet metal manufacturing in this edition. The book covers almost all the information required by a learner to master Creo Parametric. The book starts with sketching and ends at advanced topics like Sheetmetal, Surface Design, 3D Printing, MBD, and Sheet metal NC manufacturing. Some of the salient features of this book are :

## In-Depth explanation of concepts

Every new topic of this book starts with the explanation of the basic concepts. In this way, the user becomes capable of relating the things with real world.

## Topics Covered

Every chapter starts with a list of topics being covered in that chapter. In this way, the user can easy find the topic of his/her interest easily.

## Instruction through illustration

The instructions to perform any action are provided by maximum number of illustrations so that the user can perform the actions discussed in the book easily and effectively. There are about 1300 small and large illustrations that make the learning process effective.

## Tutorial point of view

At the end of concept's explanation, the tutorial make the understanding of users firm and long lasting. Almost each chapter of the book has tutorials that are real world projects. Moreover most of the tools in this book are discussed in the form of tutorials.

## Project

Free projects and exercises are provided to students for practicing.

## For Faculty

**If you are a faculty member, then you can ask for video tutorials on any of the topic, exercise, tutorial, or concept.**

## Formatting Conventions Used in the Text

All the key terms like name of button, tool, drop-down etc. are kept bold.

## Free Resources

Link to the resources used in this book are provided to the users via email. To get the resources, mail us at *cadcamcaeworks@gmail.com* with your contact information. With your contact record with us, you will be provided latest updates and informations regarding various technologies. The format to write us mail for resources is as follows:

Subject of E-mail as *Application for resources of _____ book*.
Also, given your information like
*Name:*
*Course pursuing/Profession:*
*Contact Address:*
*E-mail ID:*

Note: We respect your privacy and value it. If you do not want to give your personal informations then you can ask for resources without giving your information.

## About Authors

The author of this book, Matt Weber, has authored many books on CAD/CAM/CAE available already in market. **SolidWorks Simulation Black Books** are one of the most selling books in SolidWorks Simulation field. The author has hands on experience on almost all the CAD/CAM/CAE packages. If you have any query/doubt in any CAD/CAM/CAE package, then you can contact the author by writing at cadcamcaeworks@gmail.com

The author of this book, Gaurav Verma, has written and assisted in more than 15 titles in CAD/CAM/CAE which are already available in market. He has authored **AutoCAD Electrical Black Books** which are available in both **English** and **Russian** language. He has also written **Creo Manufacturing 4.0 Black Book** which covers Expert Machinist module of Creo Parametric. He has provided consultant services to many industries in US, Greece, Canada, and UK.

## For Any query or suggestion

If you have any query or suggestion, please let us know by mailing us on ***cadcamcaeworks@gmail.com***. Your valuable constructive suggestions will be incorporated in our books and your name will be addressed in special thanks area of our books on your confirmation.

# Chapter 1

# Starting with Creo Parametric

## Topics Covered

The major topics covered in this chapter are:

- *Installing Creo Parametric 6.0.*
- *Starting Creo Parametric.*
- *Starting a new document.*
- *Terminology used in Creo Parametric.*
- *Opening a document.*
- *Closing documents.*
- *Basic Settings for Creo Parametric*
- *Model Properties in Creo Parametric*
- *Keyboard shortcuts and Mouse Functions*

## INSTALLING CREO PARAMETRIC 6.0

- If you are installing using the CD/DVD provided by PTC then go to the folder containing **setup.exe** file and then right click on **setup.exe** in the folder. A shortcut menu is displayed on the screen; refer to Figure-1.

*Figure-1. Shortcut menu for installation*

- Click on the **Run as administrator** option from the shortcut menu. The **User Account Control** dialog box will be displayed.
- Click on the **Yes** button from the dialog box. The **PTC Installation Assistant** dialog box will be displayed; refer to Figure-2.
- Click on the **Next** button from the dialog box. The **License Agreement** page will be displayed. Accept the agreement and select the check box below the agreement.
- Click on the **Next** button from the dialog box. The license page will be displayed; refer to Figure-3.

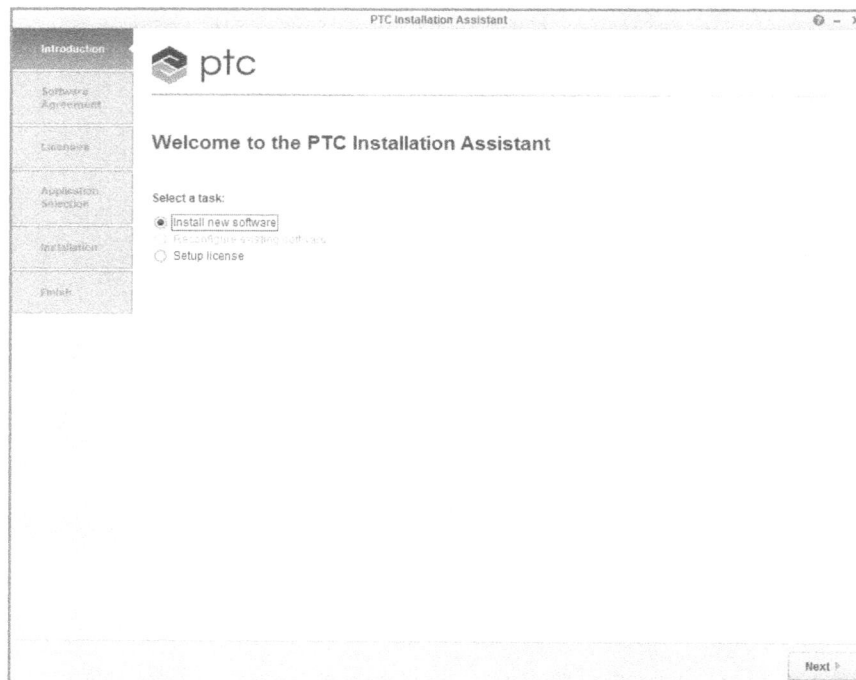

*Figure-2. PTC Installation Assistant dialog box*

*Figure-3. License page of PTC Installation*

- Enter the license file location in the edit box under **Source** column in the **License Summary** node; refer to Figure-4.

*Figure-4. License file location*

- Click on the **Next** button from the dialog box. Select the check boxes for the applications that you are licensed to install and click on the **Install** button. The software will be installed.
- After the installation is finished, click on the **Finish** button to exit.

## STARTING CREO PARAMETRIC

- Click on the **Start** button from the **Taskbar** to open the **Start Menu**.
- Click on the **All Programs** option and then click on the **PTC** folder. Next, click on the **PTC Creo Parametric 6.0** option from the menu in Windows 7. In Windows 10, click on the Start button and type Creo Parametric 6.0, the program will be displayed in the menu; refer to Figure-5.

*Figure-5. Start menu*

- Click on the program name to start application. On starting the application, the initial screen of the application is displayed as shown in Figure-6.

*Figure-6. Initial screen of Creo Parametric*

- The elements of interface of the **Creo Parametric 6.0** are shown in Figure-7.

*Figure-7. Creo Parametric interface*

The elements in the interface are discussed next.

## CREO PARAMETRIC 6.0 INTERFACE

The interface of Creo Parametric comprises of various elements like Navigator panel, Browser, Ribbon, File menu and so on. These elements are discussed next.

## Navigator panel

The **Navigator** panel is used to navigate through folders and features. This panel has three tabs named: **Model Tree**, **Folder Browser**, and **Favorites**. The **Model Tree** tab will be active only when you have opened a model/assembly file; refer to Figure-8.

*Figure-8. Model Tree*

## Model Tree

- The **Model Tree** contains all the features that are created automatically or manually while creating model or assembly.
- The green line in the **Model Tree** represents the position where the next feature will be inserted. You can drag this line up or down to create or suppress features.

- Click on the **Settings** button ⊤| ▾ to display the menu related to settings; refer to Figure-9.

*Figure-9. Model Tree settings*

- Using the **Import Settings File** option, you can open an earlier save setting file.
- To save the existing settings in a file, click on the **Export Settings File** option. The settings will be saved in .cfg file.
- Click on the **Tree Filters** option to add or remove the entities being displayed. The **Model Tree Items** dialog box will be displayed; refer to Figure-10.

*Figure-10. Model Tree Items dialog box*

- Select the check boxes that you want to display in the **Model Tree** and click on the **OK** button from the dialog box.

## Folder Browser

- Click on the **Folder Browser** button to display the **Folder Browser**; refer to Figure-11.

*Figure-11. Folder Browser*

- Using the **Folder Browser**, you can quickly browse through the folders to get the desired files.

## Favorites Tab

- The **Favorites** tab is used to display the locations saved as favorite.
- To display or hide the **Navigation** panel, click on the **Show Navigator** toggle button in the Status bar; refer to Figure-12.

*Figure-12. Show Navigator button*

# RIBBON

**Ribbon** is the area of the application window that holds all the tools for designing and editing; refer to Figure-13.

*Figure-13. Ribbon*

The tools in the **Ribbon** will be discussed later in the book as per their need in designing.

## FILE MENU

The options in the **File** menu are used to perform operations related to files; refer to Figure-14.

*Figure-14. File Menu*

The options in the **File** menu are discussed next.

## New Tool

The **New** tool is also available in the **Ribbon** and **Quick Access Toolbar**; refer to Figure-15. This tool is used to create a new file. The procedure to use this tool is given next.

*Figure-15. New Tool*

• Click on the **New** tool from the desired location. The **New** dialog box will be displayed; refer to Figure-16.

*Figure-16. New dialog box*

- Select the desired radio button from the **Type** area of the dialog box. The related sub-types will be displayed in the **Sub-type** area.
- Select the desired sub-type and specify the name in the **Name** edit box.
- For most of the file types, the **Use default template** check box is displayed. If this check box is selected then the default template is used for creating new files. The units for default templates are in **IPS**.
- Clear this check box if you want to select any other template.
- Click on the **OK** button from the dialog box. The file will be created and environment related to selected type will be displayed.
- If you have cleared the **Use default template** check box then the **New File Options** dialog box will be displayed; refer to Figure-17.

*Figure-17. New File Options dialog box*

- Select the desired template and click on the **OK** button to create the file.

The description of various file types displayed in the **New** dialog box is given next.

## Layout

The **Layout** radio button in the **New** dialog box is used to create new layout files. Layout files are used to create layouts based on the ideas. In the layout environment, you can create 2D sketches, import images, import and edit other layouts, and so on.

## Sketch

The **Sketch** radio button is used to activate sketching environment where you can create sketches.

## Part

The **Part** radio button is used to create 3D models. There are four sub-types for this radio button; **Solid**, **Sheetmetal**, **Bulk**, and **Harness**.

- The **Solid** sub-type is used to create solid/surface 3D models.
- The **Sheetmetal** sub-type is used to create sheetmetal models.
- The **Bulk** sub-type is used to create representation of model elements in assembly. For example, you need a generator of specific kind in your assembly but you do not want to model it, then you can save generator in the bill of material (BOM) by making it a bulk part.
- The **Harness** sub-type is used to creating wiring in the model. This radio button is active when you are working in an assembly and from within the assembly, you need wire harness or pipe harness.

## Assembly

The **Assembly** radio button is used to create assemblies of solid/sheet metal parts. There are 10 sub-types for this radio button; **Design**, **Interchange**, **Process Plan**, **NC Model**, **Mold layout**, **External simplified representation**, **Configurable Module**, **Configurable product**, **ECAD**, and **Instrumented**.

- The **Design** sub-type is used to create assembly designs.
- The **Interchange** sub-type is used to create a special type of assembly that you can create and then use in a design assembly. An interchange assembly consists of models that are related either by function or representation. You can create both functional interchanges (to replace functionally equivalent components) and simplify interchanges (to substitute components in a simplified representation) in the same interchange assembly. Interchange assemblies, like family tables and layouts, provide a powerful method of automatic replacement.
- The **Process Plan** sub-type is used to document the process of assembling/dis-assembling parts in the assembly.
- The **NC Model** sub-type is used to create manufacturing model of the assembly.
- The **Mold layout** sub-type is used to create layouts for molds.
- The **External simplified representation** sub-type is used to create representations of the assembly. These representations can be saved independent of the parent assembly file.
- The **Configurable module** sub-type is used to create modules that consist of parts and assemblies interrelated to each other by some relation. The module can be configured with the help of variables.

- The **Configurable product** sub-type is used to create configurable assemblies.
- The **ECAD** sub-type is used to create electronic board design.
- The **Instrumented** sub-type is used to create instrumented assembly that maintains one-way associativity of parts in the assembly. In instrumented assembly, if you modify a part outside the assembly environment then it will be updated in Instrumented assembly. But, if you modify part in assembly environment then it will not be updated at source.

## Manufacturing

The **Manufacturing** radio button is used to create manufacturing models for generating CNC codes. There are 9 sub-types available for this radio button; **NC assembly**, **Expert Machinist**, **CMM**, **Sheetmetal**, **Cast cavity**, **Mold cavity**, **Harness**, **Process plan**, and **Additive Manufacturing**.

- The **NC assembly** sub-type is used to create assembly model for manufacturing a component on nc machine like turning centers and milling centers.
- The **Expert Machinist** sub-type is used to perform the same work as done in NC assembly sub-type. But, the Expert Machinist is a lite version of NC assembly and is having limited tools to perform action. The **Creo Manufacturing 4.0 Black Book** is already available in the market on this topic from same author.
- The **CMM** sub-type is used to create programs for Coordinate Measuring Machines.
- The **Sheetmetal** sub-type is used to generate NC codes for sheet metal forming machines.
- The **Cast cavity** sub-type is used to create cavity for casting the model part.
- The **Mold cavity** sub-type is used to create mold for the model part.
- The **Harness** sub-type is used to create wiring/piping for manufacturing/lay outing.
- The **Process plan** sub-type is used to generate manufacturing process plan.
- The **Additive Manufacturing** sub-type is used to create products by 3D printing and other additive manufacturing techniques.

## Drawing

The **Drawing** radio button is used to create 2D drawings from models and assemblies. The drawings are the paper representations of models and assemblies for manufacturing.

## Format

The **Format** radio button is used to create format for drawings. These formats are later used in multiple drawings as templates.

## Notebook

The **Notebook** radio button is used to create notes of ideas regarding the design.

## Open Tool

The **Open** tool is used to open Creo Parametric files and import files of other modeling applications. The procedure to use this tool is given next.

- Click on the **Open** tool from the **File** menu, **Ribbon**, or **Quick Access Bar**. The **File Open** dialog box will be displayed; refer to Figure-18.

*Figure-18. File Open dialog box*

- The left pane of the dialog box is used to access the common folders. Note that the files that are not saved yet but are in current session can be accessed by click on the **In Session** link from the **Common Folders** area in the left pane.
- Click on the **Views** tool in the toolbar of the dialog box. The drop-down will be displayed as shown in Figure-19.

*Figure-19. Views drop-down*

- Select the desired radio button to display the files in the desired manner.
- Click on the **Tools** option in Toolbar to modify parameters related to **File Open** dialog box. The drop-down will be displayed as shown in Figure-20.

*Figure-20. Tools drop-down*

- Click on the **Address Default** option from the drop-down to set default directory which will open automatically in **File Open** dialog box. The **'Look In' Default** dialog box will be displayed; refer to Figure-21.

Figure-21. 'Look In' Default dialog box

- Select the desired option from the drop-down and click on the **OK** button.
- Select the desired option from the **Sort By** cascading menu of the **Tools** drop-down to define how files will be sorted.
- Click on the **Up One Level** tool from the drop-down to open parent folder of the current file location.
- Click on the **Add to Favorites** option from the drop-down to add current folder in Favorites list. Once, you have added the folders to favorite location, click on the **Favorites** option from the **Common Folders** area at the left in the **File Open** dialog box to access them.
- If you want to remove a folder from Favorites then select it from Favorites location and click on the **Remove from Favorites** tool in the **Tools** drop-down.
- Select the **All Versions** tool from the drop-down to check all the versions of files.
- Click in the **Type** drop-down to specify the file type; refer to Figure-22.

Figure-22. Type drop-down

- Double-click on the file under the **Name** column of the dialog box. The file will open.

## Save Tool

The **Save** tool in the **File** menu is used to save the files. If you are saving a file for the first time then the **Save Object** dialog box will be displayed otherwise the file will be save automatically. The procedure to use this tool is given next.

- Click on the **Save** tool from the **File** menu, **Ribbon**, or **Quick Access Toolbar**. If you are saving a new file for the first time then the **Save Object** dialog box will be displayed; refer to Figure-23.

*Figure-23. Save Object dialog box*

- Specify the location of the file and click on the **OK** button from the dialog box to save it.
- Now, if you again click on the **Save** button then the **Save Object** dialog box will not be displayed and the file will be automatically saved with same name and settings.

## Save As

There are three options in the **File** menu for **Save As** tool; refer to Figure-24. The **Save a Copy** option is used to create a copy of the file with new name. The **Save a Backup** tool is used to create a backup copy of the current file in the same directory. The **Mirror Part** tool is used to save a mirror copy of the current file. Note that the files saved by this tool automatically synchronize with the original file.

*Figure-24. Save As options*

The procedure to use this tool is given next.

- Click on the desired tool from the **Save As** cascading menu; refer to Figure-24. (**Mirror Part** in our case).
- The related dialog box will be displayed; refer to Figure-25.

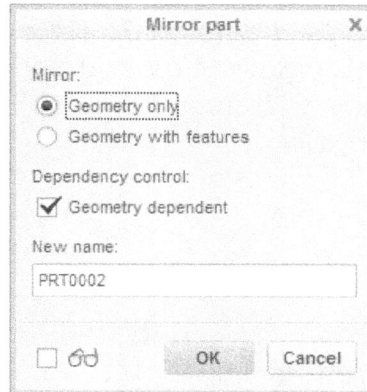

*Figure-25. Mirror Part dialog box*

- Specify the new name in the **New name** edit box and set the parameters as required.
- Click on the **OK** button from the dialog box. The part will be saved with the new name.

## Print

The **Print** tool is used to take printouts of the drawings/models for shop floor use. There are three options for printing; refer to Figure-26. The **Print** option is used to take print out of the object currently displayed in the drawing area. In this case, you can configure the printer as per your requirement. The **Quick Print** option is used to take quick prints of the objects. In this case, you can only select a printer. You cannot change the settings of the printer. The **Quick Drawing** option is displayed for models and assemblies. Using this option, you can quickly create drawings from the model/assembly and take print outs. The **Prepare for 3D Printing** option is used to prepare the model 3D Printing. You will learn more about this option later in the book. The **Order 3D Print** option is used to order the 3D printed model of current model/assembly online. The procedures to use these tools are given next.

*Figure-26. Print options*

## Print Tool

- Click on the **Print** option from the **Print** cascading menu; refer to Figure-26. The **Print** dialog box will be displayed; refer to Figure-27.

- Click on the ⁺↓ button and add the desired printer.
- Click on the **Page** tab from the dialog box and specify the desired parameters in the dialog box displayed; refer to Figure-28.

*Figure-27. Print Configuration dialog box*

*Figure-28. Page tab of Print Configuration dialog box*

- Click on the **OK** button from the dialog box to print the object.

## Quick Print tool

- Click on the **Quick Print** tool from the **Print** cascading menu. The **Print** dialog box will be displayed as shown in Figure-29.

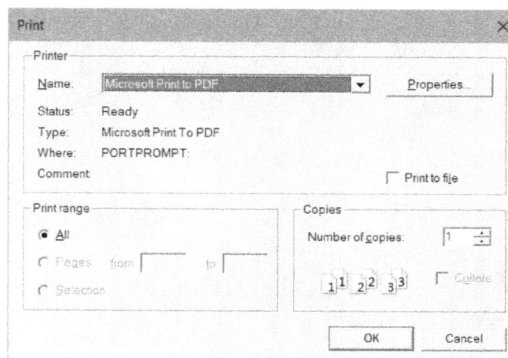

*Figure-29. Print dialog box*

- Select the desired printer from the **Name** drop-down in the dialog box.
- Set the desired parameters and then click on the **OK** button from the dialog box to print the objects.

## Quick Drawing tool

- Click on the **Quick Drawing** tool from the **Print** cascading menu. The **Quick Drawing** dialog box will be displayed; refer to Figure-30.

*Figure-30. Quick Drawing dialog box*

- Select the desired views from the **View Layout** area of the dialog box.
- You will learn more about the parameters later in the book. For the moment, click on the **OK** button from the dialog box. The **Print** dialog box will be displayed as discussed earlier.
- Click on the **OK** button from the dialog box to print.

You will learn about 3D Printing options later in the book.

## Close Tool ⌧

The **Close** tool is used to close the current file without closing the application.

## Manage File

The **Manage File** options are used to manage files of Creo Parametric. There are five options available in this cascading menu; **Rename**, **Delete Old Versions**, **Delete All Versions**, **Declare**, and **Instance Accelerator**; refer to Figure-31.

*Figure-31. Manage File options*

- The **Rename** tool in the **Manage File** cascading menu is used to rename the current file. To rename the file, click on this tool. The **Rename** dialog box will be displayed; refer to Figure-32. Specify the desired name and click on the **OK** button from the dialog box.

*Figure-32. Rename dialog box*

- The **Delete Old Versions** tool is used to delete all the previous versions of the file but to keep the latest version of the file. Note that whenever you save the file in Creo Parametric, it creates a new version of file so that anytime you can go back to previous versions. But more versions take more space on disk. So, the **Delete Old Versions** tool is used to delete the older versions. To delete the older versions of the file, click on the **Delete Old Versions** tool from the **File** menu. The **Delete Old Versions** dialog box will be displayed. Click on the **Yes** button from the dialog box to delete the older versions of the file.

- The **Delete All Versions** tool is used to delete the file and its history from session as well as working directory. This tool works in the same way as the previous tool works but it also removes the current version of file.

- The **Declare** tool is used to declare any axis or datum as global for assembly purpose. For using this tool, you must have a notebook active in session. You can create a notebook by using the **New** dialog box. Once you have declared a datum as global, each part of assembly can use it for locating itself.

- The **Instance Accelerator** tool is used to used to delete all the useless data of the current file so that the work can be done more efficiently on the file. Using this tool, you can delete the previous versions of the file that are of no use now.

On selecting this tool, the **Instance Accelerator** dialog box will be displayed; refer to Figure-33. Click on the **Purge** button to accelerate the system processing.

*Figure-33. Instance Accelerator dialog box*

# Model Properties

The **Model Properties** option is available in the **Prepare** cascading menu of the **File** menu; refer to Figure-34. This tool is used to change the properties of the model. The procedure to do so is given next.

*Figure-34. Model Properties option*

- Click on the **Model Properties** option from the **Prepare** cascading menu. The **Model Properties** dialog box will be displayed; refer to Figure-35.

# Material Selection

Using the options available in the **Materials** area of the dialog box, you can change the properties of the model related to material. The steps are given next.

- Click on the **Change** link button next to **Material** in the **Materials** area of the dialog box; refer to Figure-36.

*Figure-35. Model Properties dialog box*

*Figure-36. Materials dialog box*

- Open the desired directory and double-click on the desired material from the list of materials to add it to the list of materials used for the current model. The material will be added in the list at the bottom as shown in Figure-37.

- If you want to edit the parameters of the material before applying it to model, then double-click on the material in the **Materials in Model** area of the dialog box. The **Material Definition** dialog box will be displayed; refer to Figure-38.

*Figure-37. Material to be applied to model*

*Figure-38. Material Definition dialog box*

- The dialog box allows you to change the mechanical and thermal properties by using the **Structural** and **Thermal** tabs, respectively. Similarly, you can change the **Miscellaneous** properties and appearance of the material.
- After setting the desired properties, click on the **OK** button from the dialog box.
- Make sure the desired material is selected in the **Materials in Model** area of the dialog box as assigned; refer to Figure-39 and then click on the **OK** button from the dialog box.
- If the desired material is not assigned the right-click on the desired material from the **Materials in Model** area. A shortcut menu will be displayed; refer to Figure-40.

*Figure-39. Material assigned*

*Figure-40. Shortcut menu for material*

- Click on the **Assign** option from the menu to assign the material and then click on the **OK** button from **Question** dialog box displayed. Click on the **OK** button from the **Materials** dialog box.

## Unit System

The **Units** option in the **Model Properties** dialog box is used to change the unit system for modeling. The procedure to change the unit system is given next.

- Click on the **Change** link button for **Units** in the **Materials** area of the dialog box. The **Units Manager** dialog box will be displayed; refer to Figure-41.

*Figure-41. Units Manager dialog box*

- Select the desired unit system from the list in the dialog box and click on the **Set** button to apply the units. The **Changing Model Units** dialog box will be displayed; refer to Figure-42.

*Figure-42. Changing Model Units dialog box*

- If by mistake you have designed your model in inches or mm and you want the system to interpret 1" as 1 mm or 1 mm as 1", then click on the **Interpret dimensions** radio button.
- If you want to convert your model unit system from inch to mm or vice-versa, then click on the **Convert dimensions** radio button.
- After selecting the desired radio button, click on the **OK** button from the dialog box. The selected unit system will be applied for the model.

You can also create a unit system as per your requirement. The steps are given next.

- Click on the **New** button from the **Unit Manager** dialog box. The **System of Units Definition** dialog box will be displayed; refer to Figure-43.

*Figure-43. System of Units Definition dialog box*

- Click on the edit box of **Name** area and specify the desired name in it.
- Select the desired radio button from the **Type** area of the dialog box.
- From the **Units** area, select the units for each property.
- Click on the **OK** button from the dialog box. The unit system will be created with the specified name.
- You can set the newly created unit system as discussed earlier.

## Mass Properties

Using the **Mass Properties** option, you can change the density, inertia, and other mass properties of the model. The procedure to change the mass properties is given next.

- Click on the **Change** button next to **Mass Properties** in the **Model Properties** dialog box. The **Mass Properties** dialog box will be displayed; refer to Figure-44.

*Figure-44. Mass Properties dialog box*

- Specify the desired value for density in the **Density** edit box.
- Click on the **Calculate** button. Based on the specified density, the other properties are calculated and displayed in the dialog box.
- You can generate report of the properties by using the **Generate Report** button.
- Now, close the dialog box by clicking on the **Close** button.

You will learn about other options of the **Model Properties** dialog box and **ModelCHECK** options in **Prepare** cascading menu of the **File** menu later in the book.

## Send Options

The **Send** options in the **File** menu are used to share the model files with peers. There are two options available for sending a file to peers; **Mail Recipient (As Attachment)** and **Mail Recipient (As Link)**. The procedure for both the options is similar. The procedure is given next.

- Click on the **Mail Recipient (As Attachment)** option from the **Send** cascading menu in the **File** menu; refer to Figure-45. The **Send As Attachment** dialog box will be displayed; refer to Figure-46.

*Figure-45. Mail Recipient option*

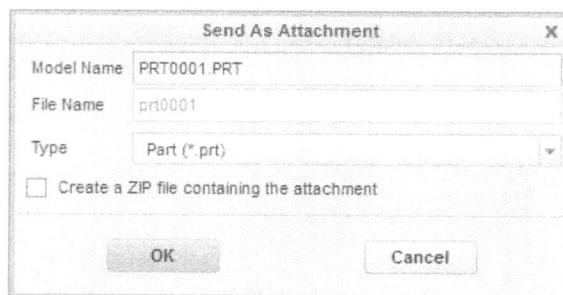

*Figure-46. Send As Attachment dialog box*

- Select the **Create a ZIP file containing the attachment** check box if you want to compress the file before sending it as attachment.
- Click on the **OK** button from the dialog box to send it. The default messenger will be displayed.
- Type the e-mail address of your peer and send it.

## Manage Session

The tools in the **Manage Session** cascading menu are used to manage session of modeling. There are about 10 tools available in the **Manage Session** cascading menu; refer to Figure-47. Some of the options are discussed next.

## Erase

The **Erase** options are used to remove the files saved in the temporary memory. Any file that is not saved by using the **Save** button are saved in the temporary memory while working. We need to free up this memory for good performance of Creo. There are three options available in **Manage Session** cascading menu for erasing session data; **Erase Current**, **Erase Not Displayed**, and **Erase Unused Model Reps**. The procedure for all the three tools is similar; click on the tool and then click on the **OK** button from the dialog box displayed.

**Erase Current**: The **Erase Current** option is used to erase the current part files from the memory.

**Erase Not Displayed**: The **Erase Not Displayed** option is used to erase the components that are not displaying but are in the session.

**Erase Unused Model Reps**: The **Erase Unused Model Reps** option is used to erase the representation of the model (part/assembly). You will learn more about representations later in the book.

*Figure-47. Manage Session cascading menu*

## Select Working Directory

The **Select Working Directory** option in the **Manage Session** cascading menu is used to set the desired directory as working directory. The procedure is given next.

- Click on the **Select Working Directory** option from the **Manage Session** cascading menu of the **File** menu. The **Select Working Directory** dialog box will be displayed; refer to Figure-48.
- Select the desired directory icon and click on the **OK** button from the dialog box. The selected directory will be set as working directory.

*Figure-48. Select Working Directory dialog box*

## Options

The **Options** button in the **File** menu is used to access the options to configure Creo Parametric. There are lots of options for configuration but we will discuss some important options only. The procedure to use the options is given next.

- Click on the **Options** button from the **File** menu. The **Creo Parametric Options** dialog box will be displayed; refer to Figure-49.

*Figure-49. PTC Creo Parametric Options dialog box*

- Click on the **Model Display** option from left area of the dialog box. The dialog box will be displayed as shown in Figure-50.

*Figure-50. Model Display page of Creo Parametric Options dialog box*

- Click on the drop-down for **Default model orientation** and select the desired orientation. Some of the designers prefer to use **Isometric** as default orientation.
- Set the shade quality to desired level by using the spinner in the **Shaded model display settings** area of the dialog box.
- Click on the **Sketcher** option from the left pane of dialog box. The options related to sketching environment will be displayed.
- Select the **Lock user defined dimensions** check box and select the **Make the sketching plane parallel to the screen** check box. The **Lock user defined dimensions** check box forces the dimensions to be locked after the user input and you cannot change the dimensions by dragging or by applying relations. The **Make the sketching plane parallel to the screen** check box is used to make the sketching plane parallel to the screen when you enter the sketching environment from modeling environment.
- Click on the **System Appearance** option from the left pane of dialog box and expand the **Graphics** node of **Global Colors** options; refer to Figure-51. Click on the **Background** button and select the white color.
- You will learn about the other options later in the book. Click on the **OK** button. A dialog box will be displayed asking you whether to store the specified settings in a configuration file. Click on the **Yes** button from the dialog box and save the configuration file at desired location.

*Figure-51. Graphics Colors options*

## SWITCHING BETWEEN FILES

To switch between various model files, click on the **Windows** button in the **Quick Access Toolbar**; refer to Figure-52.

*Figure-52. Window button in the Quick Access Toolbar*

Select the part file that you want to work with.

## SHORTCUT KEYS AND MOUSE FUNCTIONS
## IN CREO PARAMETRIC

There are various ways to use tools of Creo Parametric, like you can click on the tool directly, you can use ALT+key to select a tool, you can use CTRL+key to activate a tool, or you can use right-click shortcut menu. Some of the most common shortcuts and mouse functions are discussed next.

### Keyboard Shortcuts

Press the combination of CTRL and other key to activate the respective function as shown in Figure-53.

- Regenerate    [Ctrl] [G]

- New file    [Ctrl] [N]

- Open file    [Ctrl] [O]

- Save file    [Ctrl] [S]

- Find    [Ctrl] [F]

- Delete    [Del]

- Copy    [Ctrl] [C]

- Paste    [Ctrl] [V]

- Undo    [Ctrl] [Z]

- Redo    [Ctrl] [Y]

- Repaint    [Ctrl] [R]

- Standard view    [Ctrl] [D]

*Figure-53. Keyboard Shortcuts*

## ALT Key Menu

Press **ALT** key once while working in any environment of Creo Parametric. The key tips will be displayed; refer to Figure-54. Press the desired key from keyboard to activate the respective option.

*Figure-54. ALT key tips*

## Mouse Functions

### Dynamic viewing

#### 3D Mode

Hold down the key and button. Drag the mouse.

- Spin

- Pan　Shift

- Zoom　Ctrl

- Turn　Ctrl

#### 2D Mode

- Pan

- Zoom　Ctrl

#### 2D & 3D Mode

Hold down the key and roll the mouse wheel.

- Zoom

- Fine Zoom　Shift

- Coarse Zoom　Ctrl

*Figure-55. Mouse functions*

### Mouse control

- Highlight geometry　Over geometry

- Query to next item　Until highlighted

- Select highlighted geometry

- Add or remove items from selection　Ctrl

- Construct chains or surface sets　Shift

- Clear selection　On background

*Figure-56. Mouse Control*

In this chapter, we learned about the basics of the Creo Parametric Interface. In the next chapter, we will work on the sketching environment and its function.

# SELF ASSESSMENT

1. The .......... tab in **Navigator** panel contains all the features that are created automatically or manually while creating model or assembly.

2. In Creo Parametric, .......... is the area of the application window that holds all the tools for designing and editing

3. The default unit system of Creo Parametric Modeling environment is:

(a)   IPS            (b)   MKS
(c)   CGS           (d)   FPS

4. The ............ radio button in **New** dialog box is used to create 3D models.

5. The **Process Plan** sub-type of **Assembly** type in **New** dialog box is used to document the process of assembling/disassembling parts in the assembly. (T/F)

6. The ............ radio button in **New** dialog box is used to create notes of ideas regarding the design.

7. The files that are not saved yet but are in current session can be accessed by clicking on the ............ link from the **Common Folders** area in the left pane of **File Open** dialog box.

8. The ............ option of **File** menu is used to prepare the model 3D Printing.

9. The ............ option in **Prepare** cascading menu of **File** menu is used to change the properties of the model like material, accuracy, units, and so on.

10. CTRL+G shortcut key is used to perform following operation:

(a)   New file         (b)   Repaint
(c)   Regenerate       (d)   Find

11. The **Instance Accelerator** tool is used to clear ram to speed up function of Creo Parametric. (T/F)

12. CTRL+D shortcut key is used to perform following operation:

(a)   New file         (b)   Repaint
(c)   Regenerate       (d)   Standard View

# FOR STUDENT NOTES

Answer to Self Assessment Questions:
1. Model Tree 2. Ribbon 3. (a) 4. Part 5. T 6. Notebook 7. In Session 8. Prepare for 3D Printing 9. Model Properties 10. (c) 11. F 12. D

# Chapter 2

# Sketching

## Topics Covered

The major topics covered in this chapter are:

- *Starting Sketching environment.*
- *Setting Grid and other parameters.*
- *Sketching tools*
- *Sketch editing tools*
- *Dimensioning tools*
- *Geometric Constraints*

# STARTING SKETCHING ENVIRONMENT

The Sketching environment is used to draw or manipulate sketches. The procedure to start sketching environment is given next.

- Click on the **New** button from the **File** menu, **Ribbon**, or **Quick Access Toolbar**. The **New** dialog box will be displayed; refer to Figure-1.

*Figure-1. New dialog box*

- Select the **Sketch** radio button from the dialog box and specify the name of the file in the **Name** edit box.
- Click on the **OK** button from the dialog box. The sketching environment will be displayed; refer to Figure-2.

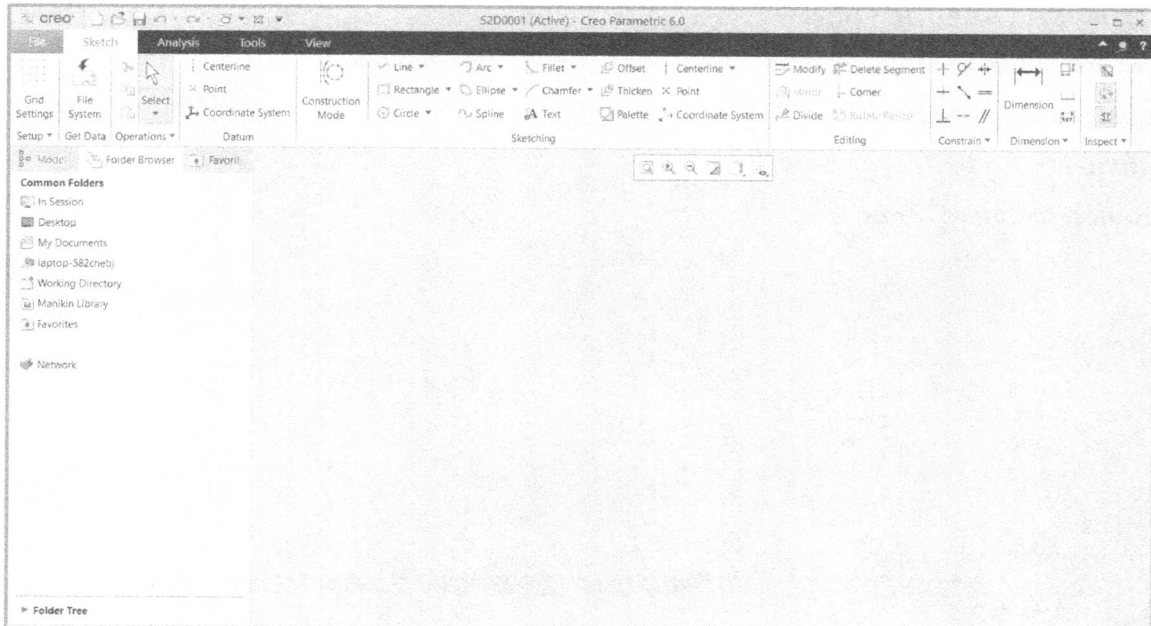

*Figure-2. Sketching environment*

# SETTING UP GRID

The Grid is used to help the designer in creating sketches. It is not always necessary to use the grid, it's more often up to the designer. The procedure to setup and display the grid is given next.

- Click on the **Grid Settings** button from the **Setup** panel in the **Ribbon**. The **Grid Settings** dialog box will be displayed; refer to Figure-3.
- Specify the desired values for the grid and then click on the **OK** button from the dialog box. The values for the grid will be saved in the configuration.
- To display the grid, click on the **Sketcher Display Filters** drop-down in the **In-Graphics** toolbar; refer to Figure-4. The list of options will be displayed.

*Figure-3. Grid Settings dialog box*

*Figure-4. Sketcher Display Filters*

- Select the **Grid Display** check box to display the grid. To hide the grid, you need to clear the check box.

# SKETCHING TOOLS

There are various sketching tools available in the **Sketching** panel of the **Sketch** tab in the **Ribbon**; refer to Figure-5.

*Figure-5. Sketching Tools*

We will discuss one by one the procedures to use the tools in the **Sketching** panel.

## Line Tools

There are two tools in Creo Parametric to create lines in the drawing; **Line Chain** and **Line Tangent**. The **Line Chain** tool is used to create a chain of lines. The procedure to create the line chains is given next.

### Line Chain tool     

- Click on the down arrow next to **Line Chain** tool. The list of the tools will be displayed as shown in Figure-6.

*Figure-6. Line Tools*

- Click on the **Line Chain** tool from the list of tools. An end point will get attached to the cursor and you will be asked to specify a point on the screen.
- Click at the desired location to specify the start point of the line. A rubber band line will get attached to the cursor.
- Stretch the line in the desired direction and click at the desired location to create the line. As soon as you specify the end point, you are asked to specify the next point.
- Keep on specifying the end points until you get the desired lines. Press the **Middle Mouse Button (MMB)** to exit the tool.

After specifying the start point of the line, if you:

- Move the cursor in horizontal direction. Then the - symbol gets attached to the line; refer to Figure-7. This - symbol designates that the line is constrained as Horizontal line. You will learn about the constraints later in the chapter.

*Figure-7. Horizontal line*

- Move the cursor in the vertical direction. Then the | symbol gets attached to the line; refer to Figure-8. Similar to the Horizontal constraint, it designates Vertical constraint.

*Figure-8. Vertical line*

## Line Tangent   Line Tangent

- Click on the **Line Tangent** tool from the **Line** drop-down in the **Ribbon**. You will be asked to select the two entities to which the line should be tangent.
- Select the two circular entities at the places where you want to specify the start point and end point of the line. Refer to Figure-9.

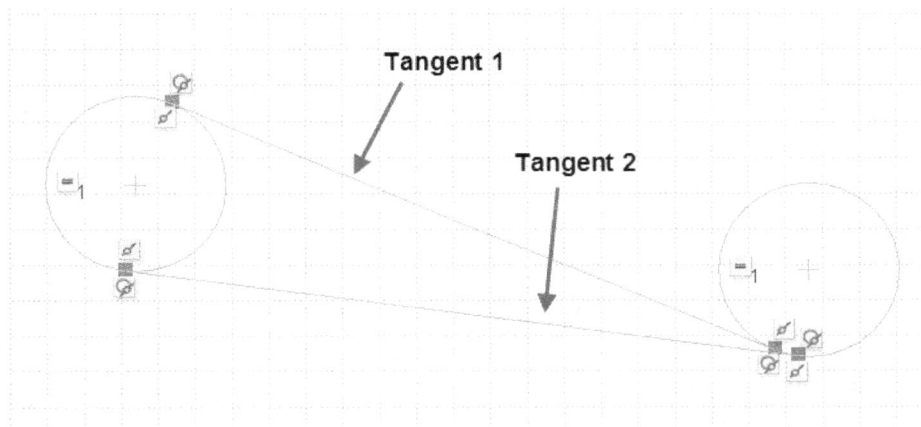

Tangent 1

Tangent 2

*Figure-9. Tangent lines*

- Note that is displayed at the contact points between circles and lines. The refers to Tangent constraint.

## Rectangle Tools

There is a list of four tools in Creo Parametric that are used to create rectangles; **Corner Rectangle**, **Slant Rectangle**, **Center Rectangle**, and **Parallelogram**. The procedure to create rectangles by using these tools is given next.

## Corner Rectangle

The **Corner Rectangle** tool is used to create rectangle by using the corner points of the rectangle. The procedure is given next.

- Click on the **Corner Rectangle** tool from the **Rectangle** drop-down; refer to Figure-10. You are asked to specify the first corner point.

*Figure-10. Corner Rectangle*

- Click in the drawing area to specify the first point of the rectangle. The other corner point will get attached to the cursor.
- Click to specify the other corner point. The rectangle will be created.
- Press the middle click to exit the tool. The rectangle will be displayed; refer to Figure-11.

*Figure-11. Rectangle created*

## Slanted Rectangle

The **Slanted Rectangle** tool is used to create rectangles aligned at some defined angle. The procedure is given next.

- Click on the **Slanted Rectangle** tool from the **Rectangle** drop-down. You are asked to specify the first point of the rectangle.
- Click to specify the first point of the rectangle. End point of a rubber-band line will get attached to the cursor.
- Click to specify the end point of the line at desired inclination. The adjacent side of the rectangle will get attached to the cursor.
- Click at the desired distance to create the rectangle.
- Press the middle mouse button to exit the tool.

## Center Rectangle

The **Center Rectangle** tool is used to create a rectangle using the center point of the rectangle and a corner point of the rectangle. The procedure to create a rectangle by this tool is given next.

- Click on the **Center Rectangle** tool from the **Rectangle** drop-down. You are asked to specify the center point of the rectangle.
- Click in the drawing area to specify the center point. The corner point of the rectangle will get attached to the cursor; refer to Figure-12.

*Figure-12. Corner point of center rectangle attached to cursor*

- Click to specify the corner point of the rectangle.

## Parallelogram

Parallelogram as the name suggests is made up of parallel sides. The procedure to create the parallelogram is given next.

- Click on the **Parallelogram** tool from the **Rectangle** drop-down. You are asked to specify the start point of baseline of the parallelogram.
- Click in the drawing area to specify the point. End point of a rubber band line will get attached to the cursor.
- Click to specify the end point of the line. Corner point of the parallelogram will get attached to the cursor.
- Click at the desired inclination to specify the corner point of the parallelogram. The parallelogram will be created.
- Press the middle mouse button to exit the tool.

    Note that now onwards, we will not write the step to exit the tool. By default, you need to press the middle button mouse to exit the tool.

## Circle Tools

The tools available in the **Circle** drop-down are used to create circle with various methods. There are four tools available in the **Circle** drop-down; **Center and Point**, **Concentric**, **3 Point**, and **3 Tangent**. The procedures to create circles by these tools are given next.

## Center and Point tool

The **Center and Point** tool is used to create circle by specifying center point and one point on its circumference. The procedure to use this tool is given next.

- Click on the **Center and Point** tool from the **Circle** drop-down; refer to Figure-13. You are asked to specify the center point of the circle.

*Figure-13. Circle drop-down*

- Click in the drawing area to specify the center point of the circle. The cursor gets attached to a point on the circumference of the circle.
- Click at the desired distance to specify the radius of the circle. The circle will be created.
- Press the middle mouse button to exit the tool.

## Concentric tool

The **Concentric** tool is used to create circle concentric to earlier created circles/arcs. The procedure is given next.

- Click on the **Concentric** tool from the **Circle** drop-down. You are asked to select an arc or circle.
- Select the arc/circle, a point on circumference of circle will get attached to the cursor with center of the circle coinciding with selected arc/circle's center; refer to Figure-14.
- Click to specify the radius of the circle.

*Figure-14. Concentric circle*

## 3 Point Circle

The **3 Point Circle** tool is used to create a circle with the help of three circumferential points. The procedure is given next.

- Click on the **3 Point Circle** tool from the **Circle** drop-down. You will be asked to specify the first circumferential point.
- Click in the drawing area to specify the first point of the circle. You will be asked to specify the second point on the circumference of circle.
- Click to specify the second point of circumference. You will be asked to specify the last point of the circumference to completely define the circle; refer to Figure-15.

*Figure-15. Three point circle*

- Click to specify the last point of the circle.

## 3 Tangent Circle

The **3 Tangent Circle** tool is used to create circle tangent to three entities as per the possibilities. The procedure to use this tool is given next.

- Click on the **3 Tangent Circle** tool from the **Circle** drop-down. You will be asked to select first entity to which you want to make the circle tangent to.
- Click on the desired entity. You will be asked to specify the second entity to be tangent; refer to Figure-16.

*Figure-16. First entity for tangency*

- Click on the second entity for tangency and then click on the next entity; refer to Figure-17.

*Figure-17. 3 Tangent Circle*

## Arcs

The tools to create arcs are available in the **Arc** drop-down. There are five tools available in the **Arc** drop-down; **3-Point/Tangent Arcs**, **Center and Ends**, **3 Tangent**, **Concentric**, and **Conic**. The procedures to use these tools are given next.

## 3-Point/Tangent Arcs

The **3-Point/Tangent Arcs** tool is used to create arc with the help of specifying three points. You can also create an arc tangent to the entity at its one of the endpoint by using this tool. The procedure is given next.

*   Click on the **3-Point/Tangent Arcs** tool from the **Arc** drop-down. You are asked to specify the start point of the arc.
*   Click in the drawing area to specify the start point. You are asked to specify the end point of the arc.
*   Click to specify the end point, you are asked to specify a point of circumference of arc to define the radius of arc; refer to Figure-18.

*Figure-18. Specifying circumferential point of arc*

*   Click to specify the circumferential point of the arc. The arc will be created.

If you want to create an arc tangent to a line/arc, then follow the steps given next.

- Click on the **3-Point/Tangent Arcs** tool from the **Arc** drop-down. You will be asked to specify the start point of arc as discussed earlier.
- Click on the end point of the arc/line to which you want to make the arc tangent to; refer to Figure-19.

*Figure-19. End point selected for tangent arc*

- Move the cursor tangentially in forward direction to the selected entity; refer to Figure-20.

*Figure-20. Cursor moved tangentially*

- Click at the desired point to specify the end point of the arc and press middle mouse button; refer to Figure-21.

*Figure-21. Arc created tangentially*

## Center and Ends

The **Center and Ends** tool, as the name suggests, is used to create an arc by specifying the center point of the arc and then the end points of the arc. The procedure to use this tool is given next.

- Click on the **Center and Ends** tool from the **Arc** drop-down. You are asked to specify the center point of the arc.
- Click to specify the center point of the arc. You are asked to specify the start point of the arc; refer to Figure-22.

*Figure-22. Start point of arc*

- Click to specify the start point. Note that the arc can be created clockwise as well as counter-clockwise.
- Click to specify the end point of the arc. The arc will be created; refer to Figure-23.

*Figure-23. Arc created*

## 3 Tangent and Concentric Arc

The **3 Tangent** and **Concentric Arc** tools work similar to **3 Tangent circle** and **Concentric circle** tools.

## Conic Arc

The **Conic** tool is used to created conical arcs. Conical arcs are used in some geometries where we design the front section of boats, aeroplanes, and automotive. The procedure to create conic arc is given next.

- Click on the **Conic** tool from the **Arc** drop-down. You are asked to specify the first point of the arc.
- Click in the drawing area to specify the first point of the arc. You are asked to specify the end point of the arc.
- Click to specify the end point of the arc. Cursor will get attached to a point on the circumference of the arc.
- Click to specify the circumferential point. On the basis of this point, the **rho** value for the conic curve will be decided. After specifying the point, the conical arc will be created; refer to Figure-24.

*Figure-24. Conical arc*

## Ellipse

There are two ways to create ellipses in Creo Parametric; by using axis ends and by using center and axis. The procedures to create ellipses by both the methods are given next.

## Axis Ends Ellipse

The **Axis Ends Ellipse** tool is used to used to create ellipses by using the axis end points. The procedure to use this tool is given next.

*   Click on the **Axis Ends Ellipse** tool from the **Ellipse** drop-down. You are asked to specify the first point of the axis for ellipse.
*   Click to specify the first point of the axis. You are asked to specify the end point of the axis.
*   Click to specify the end point. You are asked to specify the end point of the other axis; refer to Figure-25.

*Figure-25. End point to be specified for other axis*

*   Click to specify the end point. The ellipse will be created.

## Center and Axis Ellipse

Using the **Center and Axis Ellipse** tool, we can create an ellipse by specifying center and axis end point of the ellipse. The procedure to use this tool is given next.

*   Click on the **Center and Axis Ellipse** tool from the **Ellipse** drop-down. You are asked to specify the center point for the ellipse.
*   Click to specify the center point. You are asked to specify end point of the axis for ellipse; refer to Figure-26.

*Figure-26. End point selected for axis*

- Click to specify the end point. You are asked to specify the end point of the another axis; refer to Figure-27.

*Figure-27. Axis end point for creating ellipse*

- Click to specify the end point of the other axis. The ellipse will be created.

## Spline

The spline is a special geometry that can be modified to smoothen the curvature. We can use spline to design aerodynamic bodies of automotive. The procedure to create spline is given next.

- Click on the **Spline** tool from the **Ribbon**. You are asked to specify the first point of the spline.
- Click to specify the point. You are asked to specify the next point.
- Click to specify the next point and keep on specifying the points until you get the desired spline.
- To exit the tool and create the spline, press the middle mouse button twice. Refer to Figure-28.

*Figure-28. Spline created*

## Fillet

Fillet is removal of material at the sharp edges. It is used to smoothen the edges and reduce the stress at the edges. There are four tools in Creo Parametric to create fillets; `Circular`, `Circular Trim`, `Elliptical`, and `Elliptical Trim`. The procedure to create fillets by these four tools is same but the output is different. The procedure to use these tools is given next.

- Click on the desired tool from the `Fillet` drop-down. You are asked to select two entities.
- Select the first entity and then select the second entity. Note that the entities must be consecutive. The fillet will be created with random values; refer to Figure-29.

*Figure-29. Fillets created*

- Figure-29 shows the difference between four type of fillets. Note that on creating the circular fillet, a reference point is created whereas on creating the circular trim fillet, no reference point is created. Similarly, the elliptical fillet and elliptical trim fillet work.
- Press MMB to exit the tool and double-click on the value to change it. After specifying new value, press MMB to apply it.

## Chamfer

The `Chamfer` tools are used to blunt sharp edges of parts for their easy handling and more importantly their easy fitting in assembly. There are two tools in Creo Parametric for creating chamfers; `Chamfer` and `Chamfer Trim`. The procedure to use these tools is given next.

- Click on the `Chamfer` or `Chamfer Trim` tool from the `Chamfer` drop-down. You are asked to select the two entities.
- Click on the first entity and then the second entity. Note that the entity must be consecutive. On doing so the chamfer will be created; refer to Figure-30.

*Figure-30. Chamfers created.*

- Press the middle button of the mouse to exit the tool.

**Till this point, you have seen many illustrations but there are no dimensions displayed in our illustration because we want the illustrations to be neat and clean in the Book for printing. You can hide the dimensions by clearing the Dimensions Display check box from the Sketcher Display Filters drop-down in the In-Graphics toolbar; refer to Figure-31.**

*Figure-31. Dimensions Display checkbox*

# Text

This is the tool of joy for everyone. Everybody likes to write something on paper, tree, desk, and walls. Mechanical engineers like to write on parts by using machines, like brand names. **Text** tool in Creo Parametric allows us to write text on parts which are later machined by CAM. The procedure to use the **Text** tool is given next.

- Click on the **Text** tool from **Ribbon**. You are asked to specify the start point of line that will determine the height of the text.
- Click to specify the start point. You are asked to specify the end point of the line. Note that the orientation of line will determine orientation of the text.
- Click to specify the end point of the line. The **Text** dialog box will be displayed; refer to Figure-32.
- Click in the **Text** edit box and type the desired text.
- Click on the **Text Symbol** button to add symbols. The **Text Symbol** dialog box will be displayed; refer to Figure-33.
- Select the desired symbols from the dialog box and then click on the **Close** button from the dialog box to exit. The **Text** dialog box will become active again.

*Figure-32. Text dialog box*

*Figure-33. Text Symbol dialog box*

- In the **Position** area, select the horizontal and vertical justification for the text by using the **Horizontal** and **Vertical** drop-downs.
- Using the **Aspect ratio** slider and edit box, you can change the aspect ratio of the text. Aspect ratio is the ratio of width and height of the text. If you increase the aspect ratio, the text will occupy wider area; refer to Figure-34.

Aspect ratio 1.0        Aspect ratio 1.5

*Figure-34. Aspect ratio*

- Using the **Slant angle** edit box and slider, you can tilt the text at its position; refer to Figure-35.

*Figure-35. Slant angle set to 30 degree*

- The **Place along curve** check box is used to place the text along a curve. Select the **Place along curve** check box and then select the curve; refer to Figure-36. The text will be aligned along the curve.

Curve selected

*Figure-36. Text along curve*

## Offset

The **Offset** tool is used to create a copy of the selected entity at a specified distance. The procedure to create offset entities is given next.

- Click on the **Offset** tool from the **Ribbon**. The **Type** dialog box will be displayed; refer to Figure-37. There are three radio buttons available in the dialog box; **Single**, **Chain**, and **Loop**.

*Figure-37. Type dialog box*

- On selecting the **Single** radio button, you can offset a single entity. On selecting the **Chain** radio button, you can offset a chain of selected entities. On selecting the **Loop** radio button, you can offset closed loop of entities.
- Select the desired radio button from the **Type** dialog box.
- Select an entity, chain or loop of entities as required. You are asked to specify the offset distance and an arrow will be displayed with the entity, specifying the direction of offset; refer to Figure-38.
- Specify the desired distance in the edit box displayed. You can specify negative value for creating offset in opposite direction. Click on the **OK** button from the edit box. The offset entity/entities will be created; refer to Figure-39.

Enter offset in the direction of the arrow [Quit]

Edit box for offset distance

Type                    X

Select Offset Edge
● Single
○ Chain
○ Loop

Close

Offset direction

*Figure-38. Offset options*

L₁
L₁

Single Offset

Last entity of chain

//₁    //₂    //₃    //₄

First entity of chain

//₁    //₂    //₃    //₄

Loop Offset

Chain Offset

*Figure-39. Offset entities*

# Thicken

The **Thicken** tool is used to create offset copy on the both sides of the selected entity. The procedure to use this tool is similar to the **Offset** tool. The procedure is given next.

- Click on the **Thicken** tool from the **Ribbon**. The **Type** dialog box will be displayed; refer to Figure-40.

*Figure-40. Type dialog box*
*for Thicken tool*

- The radio buttons in the **Select Thicken Edge** area are discussed earlier. The radio buttons in the **End Caps** area are used to specify the condition of ends of the offset entities. The **Open** radio button is used to keep the ends of offset as open. The **Flat** radio button is used to make the ends of offset entities joined by lines. The **Circular** radio button is used to make the ends of offset entities joined by semi-circles.

- Select the desired radio buttons from the **Type** dialog box and click on the entity in the drawing area. The **Enter thickness** edit box is displayed; refer to Figure-41.

*Figure-41. Enter thickness edit box*

- Specify the desired distance between two offset copies of the selected entity and press **ENTER**. The edit box to specify distance between selected entity and created offset copies is displayed; refer to Figure-42.

*Figure-42. Offset distance edit box*

- Specify the desired distance for offset in the edit box and press ENTER. The offset entities will be created; refer to Figure-43.

*Figure-43. Thickness entities created*

## Palette

The **Palette** tool is used to import or use already created entities. The procedure to use this tool is given next.

- Click on the **Palette** tool from the **Ribbon**. The **Sketcher Palette** dialog box will be displayed; refer to Figure-44.

*Figure-44. Sketcher Palette dialog box*

- Switch between various categories by using the tabs. Double-click on the object that you want to use in the sketch. A plus sign will get attached to the cursor.
- Click in the drawing area to place the selected object. The object will be placed at the location you clicked and the **Import Section** tab is displayed; refer to Figure-45.

*Figure-45. Imported section*

- Using the Rotate handle, Move handle, and Scale handle; you can rotate, move and scale the imported section/object. To use the handles, click and hold the LMB on the handle icon, and drag the mouse to increase or decrease the value.
- You can also use the **Rotate** edit box and **Scale** edit box to specify the rotate and scale values respectively.
- After specifying the desired values, click on the **OK** button from the **Import Section** contextual tab. The section will be placed; refer to Figure-46.

*Figure-46. Imported section after placing in drawing area*

• Click on the **Close** button from the **Sketcher Palette** dialog box to exit.

## Center line

Center line are the entities in Creo Parametric that are used to define a reference for creating sketch entities. When you will be creating solid models later in the book, you will get to know the importance of center line. There are two tools to create center lines; **Centerline** and **Centerline Tangent**. The procedure to use these tools is given next.

## Centerline

• Click on the **Centerline** tool from the **Centerline** drop-down in the **Ribbon**; refer to Figure-47. You are asked to select a start point for the center line.

*Figure-47. Centerline drop-down*

• Click in the drawing area to specify the start point. You are asked to specify the end point of the centerline.
• Click at the desired location to specify the end point. Note that the location of end point will basically decide the rotation value for the center line (center lines in Creo Parametric are of infinite length).

## Centerline Tangent

The **Centerline Tangent** tool works in the same way as the **Centerline** tool works. The only difference is in the selection of entities. For **Centerline** tool, we need to specify points but for the **Centerline Tangent** tool, we need to select two curves to which the centerline will be tangent. Rest of the procedure is same; refer to Figure-48.

Second entity
selected

Centerline tangent

First entity
selected

*Figure-48. Tangent centerline*

## Point

The **Point** tool as the name suggests is used to create points in the drawing area. These points can later be used to reference other sketch entities. The procedure to create points is given next.

- Click on the **Point** tool from the **Ribbon**. You are asked to specify the position of the point.
- Click at the desired location to specify the point. You can keep on clicking at the desired location until you get the required number of points.
- Press the middle button of mouse to exit the tool.

## Coordinate System

Coordinate system is a reference for all the entities sketched in Creo. All the base dimensions are measured from coordinate system; refer to Figure-49. The procedure to place coordinate system is similar to **Point** tool. Click on the **Coordinate System** tool from the **Ribbon** and click at the desired location in the drawing area to place it.

8.74

1.82

17.19

12.73

8.86

17.21

17.19

Co-ordinate system

*Figure-49. Coordinate system placed*

## Construction Mode

Sometimes while creating the sketch entities, you may require some construction entities that can support in creating sketches. The procedure to create construction entities is given next.

- Click on the **Construction Mode** toggle button from the **Ribbon**; refer to Figure-50.

*Figure-50. Construction Mode button*

- Now, you can create desired construction entity by using the sketching tools discussed earlier. Figure-51 show some construction entities created by using sketching tools.

*Figure-51. Construction entities*

- Click on the **Construction Mode** toggle button to exit the construction mode.

There is no benefit in creating sketches without dimensions in Creo Parametric. So, we will discuss the procedures to apply dimensions to the sketches.

## DIMENSION TOOLS

Unlike other CAD packages, Creo Parametric has very compact tool box for dimensioning. There are only four tools to apply dimensions to entities; refer to Figure-52. These tools are discussed next.

*Figure-52. Dimension panel*

## Dimension tool

The **Dimension** tool is single tool with multiple functionality. Using the **Dimension** tool, you can dimension any type of object in Creo Parametric. There are no separate tools for dimensioning different type of entities like line, circle, spline, and so on. After creating sketches, you need to apply dimension to them for design purpose. The procedure to use this tool is given next.

- Click on the **Dimension** tool from the **Dimension** panel in the **Ribbon**. You will be asked to select entities to be dimensioned. We have different selection patterns for different entities to be dimensioned. These patterns are discussed next.

## Dimensioning Line

There are two ways to dimension a line.

- After selecting the **Dimension** tool, click on the line and press the middle click at the location where you want to place the dimension; refer to Figure-53. Now, you can change the value of dimension and press **ENTER**.

- After selecting the **Dimension** tool, one by one click on both the end points of the line and then place the dimension by pressing the middle mouse button; refer to Figure-53. Now, you can change the value of dimension and press **ENTER**.

*Figure-53. Line dimensioned*

## Dimensioning Circle

- After selecting the **Dimension** tool, click on the circumference of circle and press the middle button of mouse to place the dimension.
- By default, the radius dimension is applied to the circle. To convert it to diametrical dimension, select the dimension text by clicking on it and hold the right click button of mouse. A shortcut menu will be displayed; refer to Figure-54.

*Figure-54. Shortcut menu for dimensioning of circle*

- Click on the **Diameter** button from the shortcut menu. The dimension will be converted to diameter. You can do the vice-versa by following the same procedure.

## Dimensioning Arc

- After selecting the **Dimension** tool, click on the circumference of circle and press the middle button of mouse to place the dimension. The radius of arc will be specified.
- Now, you need to specify the span angle of the arc. Make sure the **Dimension** tool is selected. Click on the first point of the arc, then center of the arc, and then click on the end point of the arc.
- Press the Middle mouse button at the desired location to place the dimension. Angle of the dimensional arc will be specified; refer to Figure-55.

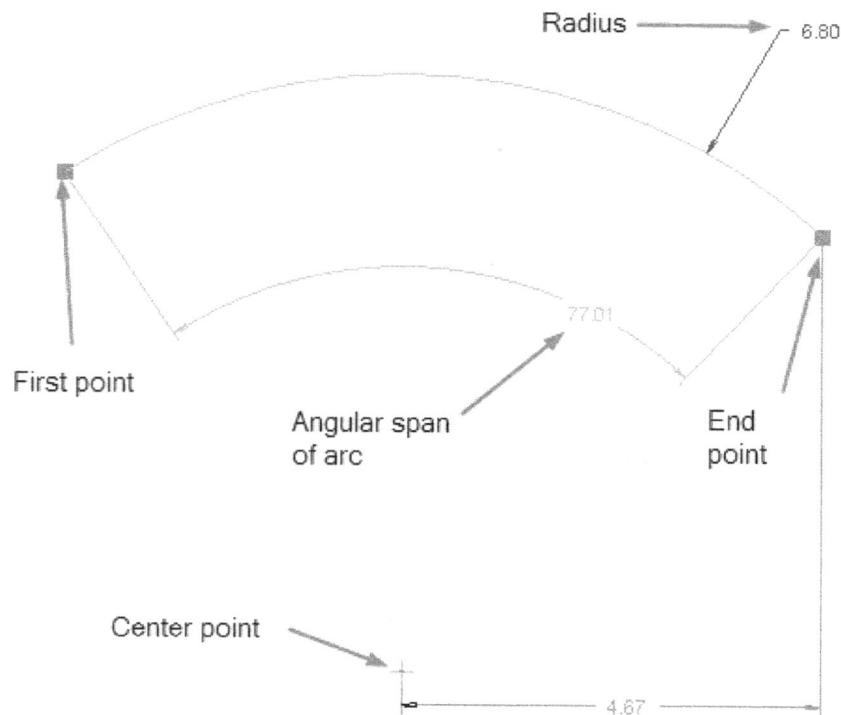

*Figure-55. Dimensioning of arc*

- You can change the angular dimension of arc to arc length. To do so, select the angle of arc and hold the right-click button of mouse. A shortcut menu will be displayed; refer to Figure-56.

*Figure-56. Convert to length option*

- Select the **Length** button from the menu. The dimension will be changed to arc length.

## Angular dimensions

One type of angular dimension was explained in dimensioning arcs. If you have two line for which you want to specify the angle, follow the steps given next.

- Click on the **Dimension** tool.
- Select the first line and then select the second line.
- Press the middle button of mouse in the region bound by two line; refer to Figure-57.

*Figure-57. Angle between lines*

- Note that the dimension value changes on the basis of location selected for placing the dimension. For example, if we press the MMB at the location outside the area bound by two lines, then the obtuse angle will be displayed; refer to Figure-58.

*Figure-58. Obtuse angle dimension*

**You can move the text of any dimension by dragging it to the desired place.**

## Dimensioning Points

Rest of the sketching entities are dimensioned with the help of their reference points. We will now discuss about dimensioning an ellipse and on the basis of that you can dimension all the other entities which are dimensioned by their points.

• Click on the **Dimension** tool if not active. Next, click one by one on the two quadrant points adjacent to each other; refer to Figure-59.

Adjacent Corner points

*Figure-59. Points of ellipse*

• Press the middle button of mouse at the desired location to place the dimension. Similarly, you can use the other points to specify the dimensions.

## Perimeter

Sometimes, we need to manipulate the perimeter of circle, rectangles and other sketched entities. For those cases, we need the perimeter dimension. The steps to create and manipulate the perimeter dimension are given next.

- Select the geometry by using window select for which you want to create the perimeter dimension and click on the **Perimeter** tool ▭ from the **Dimension** panel in the **Ribbon**. You are asked to select the varying dimension which will be manipulated to get the desired perimeter.
- Select the desired dimension (Note that the dimension selected must not be locked one). Refer to Figure-60.

*Figure-60. Perimeterical Dimension*

- Double-click on the parameter dimension and change the value as desired.

## Baseline Dimensioning

Sometimes, our sketch becomes so much complex that we are forced to remove arrows of dimension to make the sketch appear neat. In such cases, we use the Baseline dimensioning. The Baseline dimensioning is also called Ordinate dimensioning. This dimensioning also help programmers who are using NC machines for machining the parts. The procedure to dimension with baseline is given next.

- Click on **Baseline** tool from the **Dimension** panel in the **Ribbon**. You are asked to specify a reference for ordinate dimension.
- Select the desired point or line to specify the origin for ordinate dimension; refer to Figure-61.

*Figure-61. Line selected*

- Press the middle mouse button at location approximately collinear to the selected line; refer to Figure-62.

*Figure-62. Ordinate dimension*

- Now, click on the **Dimens**ion tool from the **Dimension** panel in the **Ribbon** and then click on the ordinate dimension just created.
- Next, click on the line that you want to dimension and press the middle button of mouse to place the dimension; refer to Figure-63.
- Similarly, you can specify the ordinate in vertical direction; refer to Figure-64.

*Figure-63. Ordinate dimensioning*

*Figure-64. Vertical ordinate dimension*

## Reference Dimension

In some of the sketches, the applied dimensions are enough to fully constrain the sketch but still we need some more dimensions for quality check purposes. These dimensions are called Reference dimensions. Now, we will do an activity for reference dimensioning.

- Click on the **Dimension** tool from the **Dimension** panel in the **Ribbon**. You are asked to select the reference for dimensioning.
- Create a sketch of rectangles and line as shown in Figure-65.

*Figure-65. Sketch to be dimensioned*

- Dimension the sketch as shown in Figure-66.

*Figure-66. Sketch after dimensioning*

- Now, using the **Dimens**ion tool, try to dimension the thick line. You will get the **Resolve Sketch** dialog box; refer to Figure-67.

*Figure-67. Resolve Sketch*

- This dialog box shows that the sketch is going to be over defined i.e. dimensioned more than required. Now, we have two options; either delete the extra dimension which is **4.04** in this case, or make it reference dimension.

- If you still want the dimension to be displayed for whatever purpose, then click on the **Dim>Ref** button from the **Resolve Sketch** dialog box. The dimension will be displayed as shown in Figure-68.
- You can apply this reference dimension by using the **Reference** tool in the **Dimension** panel of the **Ribbon**.

*Figure-68. Reference dimension applied*

# GEOMETRIC CONSTRAINTS

Besides the dimensions, we also constrain sketches by using the Geometric Constraints. Geometric constraints are the restrictions on the sketches like, two selected lines will be of equal length, the two selected circles will be of equal diameter, and so on. Various geometric constraints available in Creo Parametric are discussed next.

## Vertical Constrain

This constraint is used to make selected line as vertical. You can also make two points to share horizontal distance from a reference point. The procedure to use this tool is given next.

- Click on the **Vertical** button ⊥ from the **Constrain** panel in the **Ribbon**. You are asked to select entities to be constrained.
- Click on the line that you want to be vertical; refer to Figure-69. The line will be aligned vertically as shown in Figure-70.

*Figure-69. Inclined line*

*Figure-70. Vertical line created*

• Similarly, you can align two points vertically by selecting them one by one after choosing the **Vertical** button; refer to Figure-71.

*Figure-71. Points constrained vertically*

## Horizontal Constrain

This constraint is used to make selected line as horizontal. You can also make two points aligned horizontally. The procedure to use this tool is given next.

• Click on the **Vertical** button ⊥ from the **Constrain** panel in the **Ribbon**. You are asked to select entities to be constrained.
• Click to the line that you want to be horizontal; refer to Figure-72. The line will be aligned horizontally as shown in Figure-73.

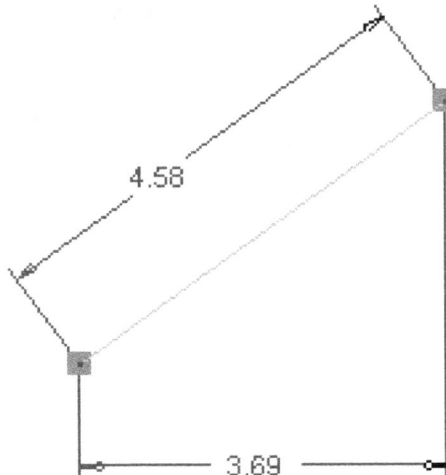

4.58

3.69

*Figure-72. Inclined line*

*Figure-73. Horizontal line created*

- Note that the length of the line is calculated as per the Pythagorus theorem.
- Similarly, you can align two points vertically by selecting them one by one after choosing the **Horizontal** button; refer to Figure-74.

*Figure-74. Points constrained horizontally*

## Perpendicular Constrain

The Perpendicular constrain is used to make a line/entity perpendicular to the other entity. The procedure to apply the constrain is given next.

- Click on the **Perpendicular** button ⊥ from the **Constrain** panel in the **Ribbon**. You are asked to select entities to be constrained.
- Select the first entity and then select the second entity. The lines will be constrained perpendicularly; refer to Figure-75.

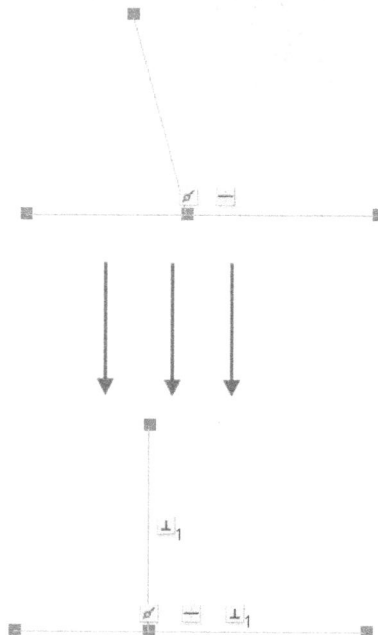

*Figure-75. Lines constrained perpendicularly*

- Similarly, you can make a arc perpendicular to the other entity; refer to Figure-76.

*Figure-76. Entities perpendiculary constrained*

## Tangent Constrain

The Tangent constrain is used to make two entities tangent to each other. The entities can be combination of line and arc, arc and arc, arc and circle, circle and circle, ellipse with arc, line, or circle and so on. The procedure to apply this constrain is given next.

- Click on the **Tangent** button ⌀ from the **Dimension** panel in the **Ribbon**. You are asked to select two entities that you want to be tangent to each other.
- Select the two entities one by one. The selected entities will become tangent to each other; refer to Figure-77.

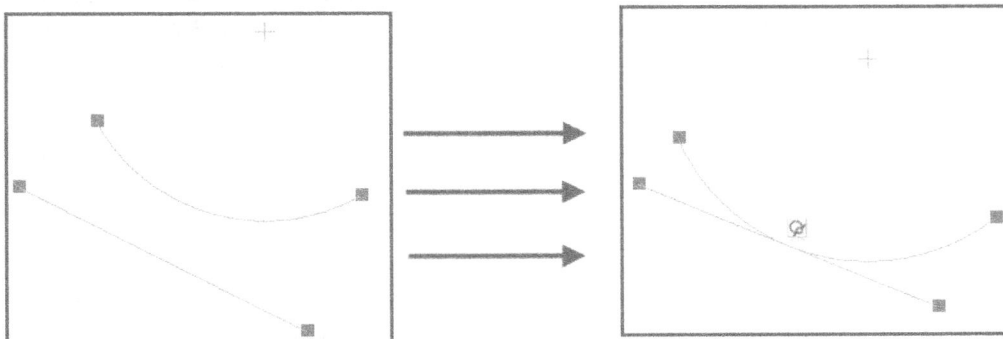

*Figure-77. Tangent entities*

## Mid-point Constrain

The **Mid-point** button is used to place the selected point at the middle of the other selected entity. The procedure to apply the Mid-point constrain using this tool is given next.

- Click on the **Mid-Point** tool from the **Dimension** panel in the **Ribbon**. You are asked to select a point.
- Select a point (it can be of any sketch entity also). Then, select the entity on whose mid-point you want to place the selected point. The point will be placed at the mid point of the selected entity; refer to Figure-78.
- In the same way, you can place selected point at the mid-point of arc, circle, ellipse and so on.

*Figure-78. Lines after mid-point constrain*

## Coincident Constrain

The Coincident constrain is used to make selected entities share the same location. The procedure to apply the coincident constrain is given next.

- Click on the **Coincident** button from the **Dimension** panel in the **Ribbon**. You are asked to select the entities to be coincident. Figure-79 shows four line with different orientations. We will make a close sketch from these lines by using this constrain.

*Figure-79. Lines for coincident constrain application*

- Select one by one points as per their group; refer to Figure-80. Like, points in Group 1 should be selected together.

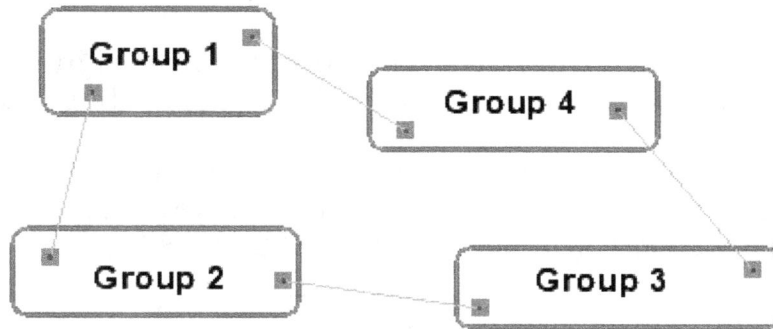

Figure-80. Groupwise points to be selected

- After selecting the points as per the group, the points will be connected as shown in Figure-81.

Figure-81. Lines after coincident constraining

- If you select two lines distant from each other and at different orientation, then the coincident constrain will make then collinear; refer to Figure-82.

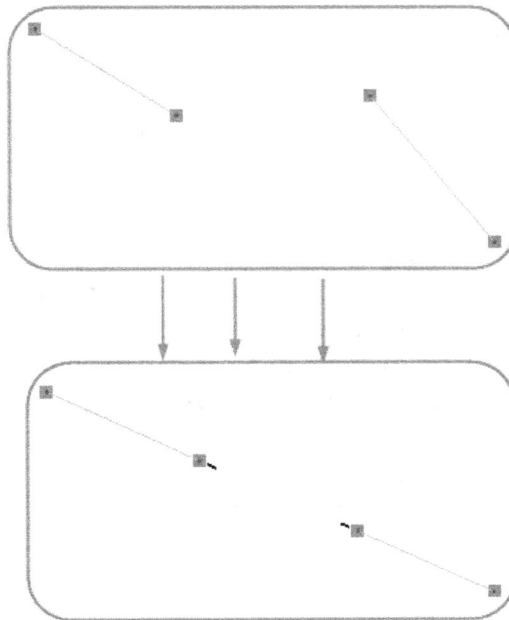

Figure-82. Lines coincident

## Symmetric Constrain

The Symmetric constrain, as the name suggests, is used to make entities symmetric about a center line. The procedure to apply Symmetric constrain is given next.

- Click on the **Symmetric** button from the **Constrain** panel in the **Ribbon**. You are asked to select a centerline and then the two vertices that you want symmetric about the centerline.
- Click on the center line. You are asked to select two vertices.
- One by one select the vertices that you want to make symmetric. On selecting the points, they will become symmetric; refer to Figure-83.

*Figure-83. Applying symmetric constrain*

- You can apply the symmetric constrain to points of any sketch entity like, circle, ellipse, spline, and so on.

## Equal Constrain

The Equal constrain is used to make dimension of two selected entities equal. Note that if one of the two entities is not constrained dimensionally then the first selected entity will be taken as reference and the other entity will be sized according to the first. The procedure to apply this constrain is given next.

- Click on the **Equal** button from the **Constrain** panel in the **Ribbon**. You are asked to select the two entities that you want to make equal.
- Click on the first entity and then click on the second entity. The entities will become equal in dimensions; refer to Figure-84.

*Figure-84. Applying equal constrain*

## Parallel Constrain

The Parallel constrain is used to make selected two lines parallel to each other. Note that if one of the two entities is not constrained dimensionally then the first selected entity will be taken as reference and the other entity will become parallel to the first entity. The procedure to apply this constrain is given next.

- Click on the **Parallel** button from the **Constrain** panel in the **Ribbon**. You are asked to select two or more line entities.
- Select the lines that you want to make parallel. The lines will be aligned parallel to each other; refer to Figure-85.

*Figure-85. Lines made parallel*

# EDITING TOOLS

Till this point, we have learned about the sketching tools, dimensioning sketches and applying geometrical constraints to the sketches. Now, we will learn to edit the sketch entities. The tools to perform editing are available in the **Editing** panel of the **Ribbon**; refer to Figure-86.

*Figure-86. Editing panel*

## Modify tool

We have earlier modified the dimensions by double-clicking on the dimension text. But there might be a scenario where you need to modify various dimension together. To modify multiple dimensions together, we use **Modify** tool. The steps to use this tool are given next.

- Click on the **Modify** tool from the **Editing** panel in the **Ribbon**. You will be asked to select the dimensions.
- One by one click on the dimensions that you want to edit. The **Modify Dimensions** dialog box will be displayed; refer to Figure-87. You can select multiple dimensions by using windows selection before activating the **Modify** tool to do the same job.

*Figure-87. Modify dimensions dialog box*

- Click in an edit box in the **Modify Dimensions** dialog box to change the dimension value. The dimension in the drawing area will be enclosed in a box; refer to Figure-88.

*Figure-88. Dimension
enclosed in box*

- Specify the desired value for the dimension in the edit box in **Modify Dimensions** dialog box. You can also use the slider next to the edit box to change the value of dimension.
- Similarly, you can change the value of other dimensions.
- Select the **Lock Scale** check box, if you want to change all the selected dimensions in same proportion. Now, move the slider or specify value for one dimension, the others will be modified automatically.
- Using the **Sensitivity** slider, you can change the sensitivity of sliders for dimension change.
- Clear the **Regenerate** check box from the dialog box.
- Click on the **OK** button from the dialog box to apply the changed dimensions.

## Mirror Tool

The **Mirror** tool is used to create a mirror copy of the selected entity. The procedure to use this tool is given next.

- Select the entity/entities whose mirror copy is to be created. The **Mirror** tool will become active.
- Click on the **Mirror** tool from the **Editing** panel. You are asked to select a center line.
- Select the center line. The mirror copy will be created; refer to Figure-89.

*Figure-89. Mirror copy*

## Divide Tool

The **Divide** tool is used to break the entity at the selected position. We will need this tool later when we will be creating 3D features. The procedure to use this tool is given next.

*   Click on the **Divide** tool from the **Editing** panel. You are asked to click on a sketch entity.
*   Click at the desired point on the entity. The entity will be broken at the clicked point; refer to Figure-90.

*Figure-90. Line divided*

## Delete Segment Tool

The **Delete Segment** tool is an intelligent eraser used to delete extra portion of the sketch entity. There are many situations in sketching when we get extra length of the sketch entity being created. In such cases, we can use this tool to remove the extra portion. The procedure to use this tool is given next.

*   Click on the **Delete Segment** tool from the **Editing** panel in the **Ribbon**. You are asked to select or drag over the entity that you want to remove.
*   Click on the entities one by one or drag over the entities that you want to delete; refer to Figure-91.

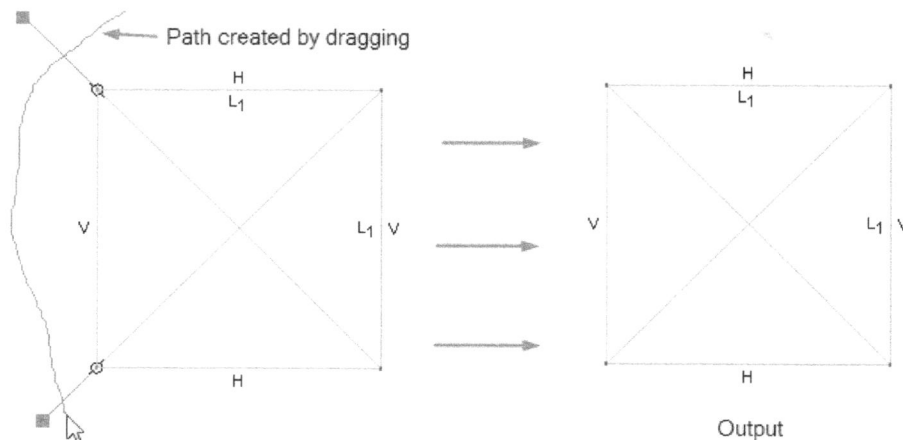

*Figure-91. Delete segment tool operation*

## Corner Tool

The **Corner** tool is used to create corner from two entities by either trimming them and/or by extending them. The procedure is given next.

- Click on the **Corner** tool from the **Editing** panel in the **Ribbon**. You are asked to select entities to be cornered.
- Select the entities which you want to trim to form corner. Refer to Figure-92.

*Figure-92. Cornered entities*

## Rotate Resize Tool

The **Rotate Resize** tool, as the name suggests, is used to rotate and re-size the selected sketched entities. The procedure to use this tool is given next.

- Select the entities that you want to rotate and re-size.
- Click on the **Rotate Resize** tool from the **Editing** panel in the **Ribbon**. The **Rotate Resize** dashboard will be displayed. Also, the drag handles will be displayed with the selected sketch entities; refer to Figure-93.

*Figure-93. Rotate Resize options*

- Click on the Move handle of the sketch and drag it to the desired location.
- Similarly, using the Rotate handle and Resize handle you can rotate and resize the sketch, respectively.
- You can also specify the desired value in the **Rotate** edit box and **Scale** edit box. In the **Rotate** edit box specify the angle value by which you want to rotate the sketch (You can specify a negative value to rotate in reverse direction). In the **Resize** edit box, specify the desired value of scale by which you want to enlarge or diminish the sketch.

In this chapter, we learned about the basic tools required for sketching. On the basis of these tools, we can draw any kind of Sketch in Creo. But there are some advanced tools in sketching which we will discuss in next chapter and we will work on practical examples and exercises.

# SELF ASSESSMENT

1. The Sketching environment is used to
(a) draw sketches        (b) manipulate sketches
(c) both                 (d) none

2. Select the .......... check box in In-Graphics Toolbar is used to display the grid.

3. The ............ tool is used to create a rectangle using the center point of the rectangle and a corner point of the rectangle.

4. The ............. tool is used to create circle using center of earlier created circles/arcs as its center point.

5. The **rho** value is required for creating conic curves. (T/F)

6. You can hide the dimensions by selecting the **Dimensions Display** check box from the **Sketcher Display Filters** drop-down in the **In-Graphics toolbar**. (T/F)

7. In case of texts in Creo Parametric, aspect ratio is the ratio of width and height of the text. (T/F)

8. The ......... tool is used to create a copy of selected sketch entities at a specified distance.

9. Which of the following tool is used to create offset copies of selected sketch entities on both sides of it?

(a) Offset       (b) Thicken
(c) Chamfer      (d) Palette

10. Using the **Dimension** tool, you can dimension any type of object in Creo Parametric. (T/F)

11. To modify multiple dimensions together, we use · · · · · · · · · · · tool.

12. The · · · · · · · · · · tool is used to trim extra portion of sketch.

13. The · · · · · · · · · · tool is used to create corner from two entities by either trimming them and/or by extending them.

## FOR STUDENT NOTES

# FOR STUDENT NOTES

**Answers to Assessment Questions:**
1. (c) 2. Disp Grid 3. Center Rectangle 4. Concentric 5. T 6. F 7. T 8. Offset 9. (b) 10.
T 11. Modify 12. Delete Segment 13. Corner

# Chapter 3

# Advanced Sketching and Practical

## Topics Covered

The major topics covered in this chapter are:

- **Sketch Inspection options.**
- **Applying relations.**
- **Practical 1**
- **Practical 2**
- **Practical 3**
- **Practical 4**
- **Practices**

# INTRODUCTION

In the previous chapter, we have learned about various sketching tools. In this chapter, we will learn about some advanced tools that help us in sketching efficiently.

# INSPECT OPTIONS

There are six tools available in the sketching environment of Creo Parametric that are used to perform inspection on sketches for their suitability in modeling. The tools are available in the **Inspect** panel; refer to Figure-1.

*Figure-1. Inspect panel*

## Overlapping Geometry

The **Overlapping Geometry** button is used to identify the overlapping sketched entities. When we will convert the sketches into 3D features there will be various instances when the sketch will be rejected. One of the possible cause for that might be overlapping geometries. So, we need to make sure that there are no overlapping geometries in the sketch. The procedure to identify the overlapping geometries is given next.

* Click on the **Overlapping Geometry** button to activate it. All the geometries that are overlapping will be displayed in red.
* Move the cursor on the entity highlighted and right click on it. A shortcut menu will be displayed; refer to Figure-2.

*Figure-2. Entity overlapping*

* Click on the **Pick From List** button. The **Pick From List** dialog box will be displayed; refer to Figure-3.

*Figure-3. Pick From List dialog box*

- Select the entity from the list that you want to delete and press the **DELETE** button from keyboard. Note that preview of selected entity is displayed in the drawing area.
- You can repeat the steps to delete all the overlapping entities.

## Highlight Open Ends

The **Highlight Open Ends** toggle button is used to highlight the open ends of the sketch by red color boxes. Later on, we will need closed sketches to create features like extruded feature, revolved features, and so on. To make sure that the sketch is closed, we will need this inspection option. The procedure to identify the open ends is given next.

- Click on the **Highlight Open Ends** toggle button to activate it. All the sketches that are open will be displayed with red boxes at their open end points; refer to Figure-4.

*Figure-4. Open ends highlighted*

## Shade Closed Loops

The **Shade Closed Loops** toggle button is used to shade the closed loops in the sketch by orange color (by default). Later on, when we will need closed sketches to create features like extruded feature, revolved features, and so on. To make sure that the sketch is closed, we will need this inspection option. The procedure to identify the open ends is given next.

- Click on the **Shade Closed Loops** toggle button to activate it. All the closed loops in the sketch will be displayed as shaded by color; refer to Figure-5.

After selecting
the button

*Figure-5. Closed loop shaded*

## Intersection Point

The **Intersection Point** button is used to identify the intersection points between the selected entities. The procedure to use this button is given next.

- Click on the **Intersection Point** button from the expanded Inspect panel. You are asked to select two entities from identifying their intersection points.
- Select one by one two entities. The intersection point will be displayed along with the **INFORMATION WINDOW**; refer to Figure-6.

INFORMATION WINDOW (entity.inf)     — □ X

File   Edit   View

ANGLE OF SLOPE ON RED ENTITY    = 143.76 deg (2.51

ANGLE OF SLOPE ON BLUE ENTITY = 82.91 deg (1.45 r

CURVATURE ON RED ENTITY    = 0.00

CURVATURE ON BLUE ENTITY = 0.00

Close

*Figure-6. Intersection points with INFORMATION WINDOW*

## Tangency Point

The **Tangency Point** button is used to identify the tangency point of a sketch entity with respect to an arc/circle. The procedure to use this tool is given next.

- Click on the **Tangency Point** button from the **Inspect** panel in the **Ribbon**. You are asked to select the two entities.
- Select the line and circle/arc. The tangency point, if possible, will be displayed along with the **INFORMATION WINDOW**; refer to Figure-7.

*Figure-7. Tangency point created*

# Entity

The **Entity** tool is used to find out all the information about the selected entity. For example, if you select a line then its length, angle and other information will be displayed. Similarly, if you select a circle, then the radius, area and other information of the circle will be displayed. The procedure to use this tool is given next.

- Click on the **Entity** tool from the expanded **Inspect** panel. You are asked to select an entity.
- Select the desired entity. The **INFORMATION WINDOW** will be displayed with the information related to the selected entity; refer to Figure-8.

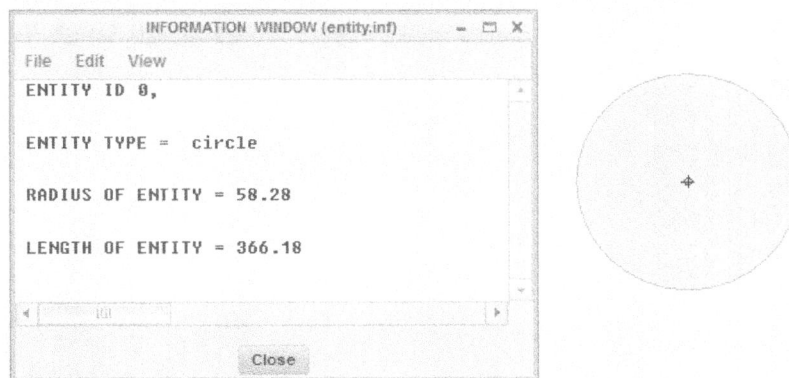

*Figure-8. Entity information*

# APPLYING RELATIONS TO DIMENSIONS

If you are working on the geometries that need mathematical relations, then you can do so by using the **Relations** tool from the **Relation** panel in the **Tools** tab of the **Ribbon**. The procedure to use this tool is given next.

- Click on the **Switch Dimensions** tool from the **Relation** panel. All the dimensions will be displayed with their designated names in place of their dimension values in the sketch; refer to Figure-9.

*Figure-9. Switching dimensions*

- Click on the **Relations** button from the **Relation** panel in the **Tools** tab of the **Ribbon**. The **Relations** dialog box will be displayed; refer to Figure-10.

*Figure-10. Relations dialog box*

- Click on the dimension for/by which you want to make the mathematical relations. For example, the diameter of bigger circle is half of the multiplication of diameters of smaller circles; refer to Figure-11.

*Figure-11. Mathematical relations in the sketch*

- The output of the formula will be as shown in Figure-12.

*Figure-12. Output of relations*

- If you try changing the value of dimension **sd2** then it will not change. But, you can change the values of **sd0** and **sd1**, and on the basis of that the value of **sd2** will automatically change.

## IMPORTING SKETCHES IN CURRENT SKETCH

The **File System** tool is used to import sketches or drawing files in the sketch you are working on. The common file types that are available for importing are *.sec, *.drw, *.igs, *.iges, *.dxf, *.dwg, and *.ai. The procedure to use this tool is given next.

- Click on the **File System** tool from the **Get Data** panel in the **Sketch** tab of the **Ribbon**. The **Open** dialog box will be displayed; refer to Figure-13.

*Figure-13. Open dialog box*

- Select the file that you want to import and click on the **Open** button from the dialog box. The sketch will get attached to cursor.
- Click at the desired location in the drawing area. The handles to modify sketch will be displayed with **Import Section** contextual tab in the **Ribbon**; refer to Figure-14.

*Figure-14. Handles on imported sketch*

- Drag the center handle of sketch section to position it at desired location.
- Using the scale and rotate handles, you can scale up/down and rotate the section. You can also specify the desired value for scale and rotation in their respective edit boxes in the **Ribbon**.
- Click on the **OK** button to create the section.

After long sessions of practicing on tools, now we will work on some drawings. Some of the practical that can be performed now are given next.

# PRACTICAL 1

In this practical, we will create a sketch for the drawing given in Figure-15.

*Figure-15. Practical 1*

## Starting Sketch

- Start Creo Parametric by double-clicking on the PTC Creo Parametric 6.0 icon from the desktop.
- Click on the **New** button from the **Quick Access Toolbar**. The **New** dialog box will be displayed.
- Select the **Sketch** radio button from the dialog box and specify the desired name in the **Name** edit box.
- Click on the **OK** button from the **New** dialog box. The sketching environment will be displayed.

## Creating outer loop

- Click on the **Line Chain** tool from the **Line** drop-down in the **Sketching** panel of the **Ribbon**. You are asked to specify the start point of the line chain.
- Create a line sketch with the shape as given in Figure-16. Note that in this figure, the dimensions are hidden. After creating this loop, we need to display the dimensions again.

*Figure-16. Sketched loop*

## Modifying Dimensions

- Select the whole sketch by box selection and then click on the **Modify** tool from the **Editing** panel. The **Modify Dimensions** dialog box will be displayed; refer to Figure-17.

*Figure-17. Modify Dimensions dialog box*

- Specify the values in the edit boxes of the dialog box as per the drawing. Make sure you have cleared the **Regenerate** check box. Note that when you click in an edit box in the dialog box, the respective dimension in the drawing is enclosed in a blue box.
- Click on the **OK** button from the dialog box to apply the dimensions.

## Drawing Circle

- Click on the **Center and Point** tool from the **Circle** drop-down. Click in the sketch at a position as shown in Figure-18.
- Double-click on the dimension values of the circle and change them as given in Figure-19.

*Figure-18. Placing circle*

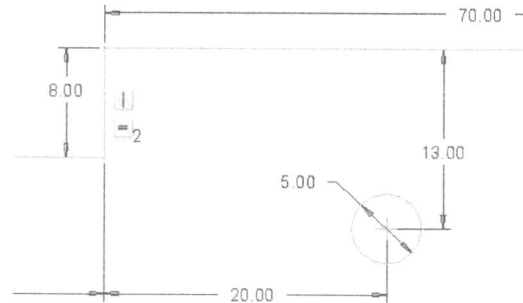

*Figure-19. Circle after specifying dimensions*

## Creating Mirror copies of Circle

- Click on the **Centerline** tool from the **Centerline** drop-down. You are asked to specify start point for the centerline.
- Click at the mid point of the vertical line and draw a horizontal center line; refer to Figure-20.

*Figure-20. Horizontal centerline created*

- Similarly, create the vertical centerline at the mid-point of longer horizontal line; refer to Figure-21.

*Figure-21. Sketch after create center lines*

## Creating Mirror Copy

- Select the circle and click on the **Mirror** tool from the **Editing** panel in the **Ribbon**. You are asked to select the center line.
- Click on the vertical center line. The mirror copy will be created; refer to Figure-22.

*Figure-22. Mirror copy of circle created*

- Similarly, create mirror copy of the two circles with respect to horizontal center line; refer to Figure-23.

*Figure-23. Sketch after creating mirror copies*

- Create the last center line and complete the sketch.

In the previous practical, we created a sketch based on the dimensioned sketch. But, in real world, we will be having an engineering drawing. So now onwards, we will create sketches from the engineering drawings.

# PRACTICAL 2

In this practical, we will create sketch from the drawing given in Figure-24.

*Figure-24. Practical 2*

## Starting Sketch

- Start Creo Parametric by double-clicking on the PTC Creo Parametric 6.0 icon from the desktop.
- Click on the **New** button from the **Quick Access Toolbar**. The **New** dialog box will be displayed.
- Select the **Sketch** radio button from the dialog box and specify the desired name in the **Name** edit box.
- Click on the **OK** button from the **New** dialog box. The sketching environment will be displayed.

## Drawing the line sketch

- Click on the **Line Chain** tool from the **Line** drop-down. You are asked to specify the start point of the line.
- Create a line sketch as shown in Figure-25.

Figure-25. Line sketch

- Click on the **Offset** tool from the **Sketching** panel. Select the **Single** radio button from the **Type** dialog box displayed.
- Click on the line with dimension value **4.50**. You are asked to specify the offset distance; refer to Figure-26.

Figure-26. Offsetting line

- Specify the offset distance as **0.495** and press ENTER. The offset copy of the line will be created; refer to Figure-27.

*Figure-27. Offseted line*

- Press the Middle button of mouse to exit the tool.
- Click on the **Line Chain** tool from the **Line** drop-down and draw a line starting from intersection point of inclined line and horizontal line to the line of length **4.50**; refer to Figure-28.

*Figure-28. Sketch after creating joining line*

- Click on the **Offset** tool from the **Sketching** panel in the **Ribbon** and select the line created recently.
- Specify the offset distance value as **0.375**. Make sure that the offset direction arrow is towards right otherwise you will have to specify **-0.375**. On doing so, an offset copy of the line will be created; refer to Figure-29.

*Figure-29. Offset copy of line created*

- Click on the horizontal line of dimension **4.50** and offset it at a distance of **0.245** downward; refer to Figure-30.

*Figure-30. Sketch after offsetting line*

## Creating Circle

- Click on the **Center and Point** tool from the **Circle** drop-down. You are asked to specify the center point for the circle.
- Draw a circle at the location similar to the one shown in Figure-31.

*Figure-31. Circle to be drawn*

- Double-click on the dimensions and change the value as shown in Figure-32.

*Figure-32. Circle after dimensioning*

- Click on the **Center and Point** tool again and create the circle with dimensions as shown in Figure-33.

*Figure-33. After creating circle*

## Trimming extra part of sketch

- Click on the **Delete Segment** tool from the **Editing** panel of the **Ribbon**. You are asked to select the entities that you want to trim.
- Delete the extra geometries from the sketch; refer to Figure-34.

*Figure-34. After trimming entities*

## Filleting

- Click on the **Circular** tool from the **Fillet** drop-down. You are asked to select two entities.
- Select the two lines as shown in Figure-35. The fillet will be created; refer to Figure-36.

*Figure-35. Lines selected for filleting*

*Figure-36. Sketch after filleting*

*   Double-click on the dimension value of the fillet and specify the value as **0.50**.

Apply rest of the dimensions as per the drawing by using the **Dimension** tool from the **Dimension** panel of the **Ribbon**.

# PRACTICAL 3
In this practical, we will create a sketch as shown in Figure-37.

*Figure-37. Practical 3*

## Starting Sketch

*   Start Creo Parametric by double-clicking on the PTC Creo Parametric 6.0 icon from the desktop.
*   Click on the **New** button from the **Quick Access Toolbar**. The **New** dialog box will be displayed.

- Select the **Sketch** radio button from the dialog box and specify the desired name in the **Name** edit box.
- Click on the **OK** button from the **New** dialog box. The sketching environment will be displayed.

## Drawing the ellipse

- Click on the **Axis Ends Ellipse** tool from the **Ellipse** drop-down. You are asked to specify the start point of the ellipse.
- Click in the drawing area to specify the start point of the ellipse. You are asked to specify the end point of the major axis of the ellipse.
- Click in the drawing area to create horizontal major axis of the ellipse; refer to Figure-38. You are asked to specify the end point for the minor axis of the ellipse; refer to Figure-39.

*Figure-38. Major axis of ellipse*

*Figure-39. Minor axis of ellipse*

- Click to specify the end point of the minor axis of ellipse and press the middle mouse button to exit the tool.
- Double-click on the dimensions of the ellipse one by one and specify the required values; refer to Figure-40.
- Create a horizontal and vertical center line from the center of the ellipse by using the **Centerline** tool.

*Figure-40. Ellipse after specifying dimensions*

## Drawing Circles

- Click on the **Center and Point** tool from the **Circle** drop-down and draw a circle of diameter **1.45** at a distance of **0.79** to the left of the center line of the ellipse.
- Similarly, draw a circle of diameter **0.33** to the right of the center line at a distance of **1.32**; refer to Figure-41.

*Figure-41. Sketch after drawing circles*

- Create a circle of radius **0.39** concentric to the circle with diameter **0.33**; refer to Figure-42.
- Click on the **Center and Ends** tool from the **Arc** drop-down and draw an arc or radius **1.90** as shown in Figure-43.

*Figure-42. Creating circle of radius 0.39*

*Figure-43. Arc created*

## Filleting the Sketch

- Click on the **Circular** tool from the **Fillet** drop-down. You are asked to select two entities for creating fillet.
- One by one select the circle and arc as shown Figure-44. Note that you need to select the arc and circle at the positions pointed by red arrows in the figure.

*Figure-44. Selection for creating fillet*

- Press the middle button of the mouse to exit the **Fillet** tool and double-click on the radius value of the fillet.
- Specify the value of the fillet as **0.25**.
- Click on the **Circular** tool from the **Fillet** drop-down and create the fillet of radius **0.13** as shown in Figure-45.

*Figure-45. Fillet of radius*

## Trimming extra geometry

- Click on the **Delete Segment** tool from the **Editing** panel in the **Ribbon**. You are asked to select the extra entities from the sketch.
- Click on the entities as shown in Figure-46.

*Figure-46. Entities selected for trimming*

- After selecting these entities, the sketch will be displayed as shown in Figure-47.

*Figure-47. Sketch after trimming*

## Mirroring entities

- Select the entities as shown in Figure-48 by holding the CTRL key from keyboard.

*Figure-48. Entities selected for mirroring*

- Click on the **Mirror** tool from the **Editing** panel. You are asked to select a center line.
- Click on the horizontal center line. The mirror copy of the selected entities will be created; refer to Figure-49. Note that dimension are hidden for clarity purpose.

*Figure-49. Creating mirror copy of entities*

## Creating Hexagon

- Click on the **Centerline** tool from the **Sketching** panel of the **Ribbon** and create a center line at the center of the circle on the left.
- Click on the **Palette** tool from the **Sketching** panel. The **Sketched Palette** dialog box will be displayed; refer to Figure-50.

*Figure-50. Sketcher Palette dialog box*

- Double-click on the **6-Sided Hexagon** option from the **Polygons** tab in the dialog box. You are asked to specify the insertion point for the hexagon.
- Click on the center of the circle having diameter of **1.45**. The hexagon will be placed near the center of the circle.

- Click on the move handle of the hexagon and drag it to the center of the circle. You may need to zoom in by scrolling down for locating the center clearly.
- Click on the **OK** button from the dashboard in the **Ribbon** and click on the **Close** button from the dialog box to exit.
- Now, we need to specify dimension of the hexagon. Click on the **Dimension** tool from the **Dimension** panel in the **Ribbon**. You are asked to select the entities.
- One by one click on the two lines shown in Figure-51 and press the middle button of mouse to place the dimension. The **Resolve Sketch** dialog box will be displayed; refer to Figure-52.

*Figure-51. Dimensioning hexagon*

*Figure-52. Resolve Sketch dialog box*

- Select the dimension that is contradicting with our newly created dimension(In our case its **sd11 = 0.16**, highlighted in box; refer to Figure-51). Click on the **Delete** button from the **Resolve Sketch** dialog box. You are asked to specify value of the created dimension.
- Specify the value as **0.79** and press ENTER from the Keyboard. We get the sketch for drawing.

# PRACTICAL 4

In this practical, we will create a sketch for the drawing given in Figure-53.

*Figure-53. Practical 4*

## Starting Sketch

- Start Creo Parametric by double-clicking on the PTC Creo Parametric 6.0 icon from the desktop.
- Click on the **New** button from the **Quick Access Toolbar**. The **New** dialog box will be displayed.
- Select the **Sketch** radio button from the dialog box and specify the desired name in the **Name** edit box.
- Click on the **OK** button from the **New** dialog box. The sketching environment will be displayed.

## Drawing Circles

- Click on the **Center and Point** tool from the **Circle** drop-down in the **Ribbon**. You are asked to specify the position of the center of circle.
- Click in the drawing area and then click to specify the size of the circle.
- Create one more circle at the same center and make it bigger than the previous one; refer to Figure-54.

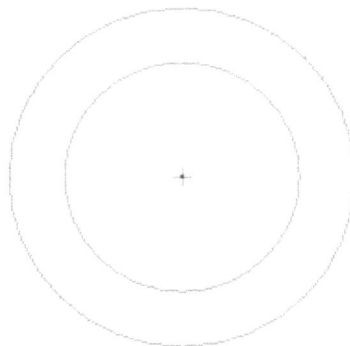

*Figure-54. Concentric circles created*

- Press the middle mouse button to exit the **Circle** tool and modify the dimension of the sketch as shown in Figure-55.

*Figure-55. Circles after dimensioning*

- Click on the **Centerline** tool from the **Sketching** panel and create a horizontal center line.
- Click on the **Center and Point** tool from the **Circle** drop-down and create the circles on the horizontal center line; refer to Figure-56. Press the middle mouse button to exit the tool.

*Figure-56. Circles created along centerline*

- Change the dimensions as shown in Figure-57.

*Figure-57. Sketch after dimensioning*

- Create smaller circles as shown in Figure-58.

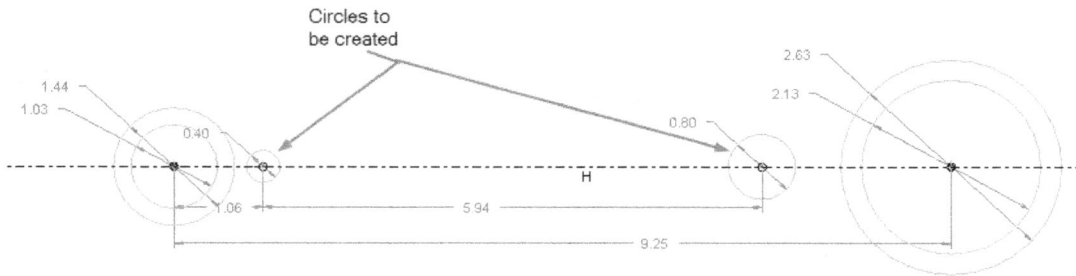

*Figure-58. Circles to be drawn*

- Create the circles as given in Figure-59. Note that the two circles are of same diameter.

*Figure-59. Circles created on the right portion*

## Creating Mirror copy of circles

- Select the two circles created in the previous step and click on the **Mirror** tool from the **Editing** panel. You are asked to select the centerline
- Click on the horizontal center line created earlier. The mirror copy of the circles will be created; refer to Figure-60.

*Figure-60. Mirror copy of circles created*

## Creating Lines

- Click on the **Line Chain** tool from the **Line** drop-down in the **Sketching** panel. You are asked to specify the start point of the line.
- Create a line connecting the two circles; refer to Figure-61.

*Figure-61. Line to be sketched*

- Select the line and create a mirror copy of the line with respect to the horizontal centerline. Specify distance between the two lines as **0.79** by using the **Dimension** tool.

## Creating Fillet

- Click on the **Circular** tool from the **Fillet** drop-down. You are asked to select two entities.

- Select the entities as shown in Figure-62 to create the fillet.

*Figure-62. Entities selected for filleting*

- Now, click on the two circles as shown in Figure-63 to create fillet between the circles.

*Figure-63. Circles to be selected*

- Similarly, create other fillets; refer to Figure-64.

*Figure-64. Sketch after creating fillets*

- Double-click on the dimension values and change the values as per the drawing; refer to Figure-65. In place of dimensioning all the fillets, you can dimension one of the similar fillets and can make all the other fillets equally constrained by using the **Equal** constraint; refer to Figure-66. Note that in Figure-66, fillet radius of 3 is denoted by E1 at all the places.

*Figure-65. After dimensioning the fillets*

*Figure-66. Fillet constrained as equal*

## Trimming the Extra geometries

- Click on the **Delete Segment** tool from the **Editing** panel. You are asked to select the segments that you want to delete.
- Select all the extra segments to delete them. The sketch after trimming will be displayed as shown in Figure-67.

*Figure-67. Sketch for Practical 4*

# PRACTICE 1

In this practice session, we will create a sketch for the drawing given in Figure-68.

*Figure-68. Practice 1*

# PRACTICE 2

In this practice session, we will create a sketch for the drawing given in Figure-69.

*Figure-69. Practice 2*

# PRACTICE 3

In this practice session, we will create a sketch for the drawing given in Figure-70.

*Figure-70. Practice 3*

# PRACTICE 4

In this practice session, we will create a sketch for the drawing given in Figure-71.

*Figure-71. Practice 4*

# PRACTICE 5

In this practice session, we will create a sketch for the drawing given in Figure-72.

*Figure-72. Practice 5*

# PRACTICE 6

In this practice session, we will create a sketch for the drawing given in Figure-73.

*Figure-73. Practice 6*

# SELF ASSESSMENT

1. The tools in .......... panel are used to perform inspection on sketches for their suitability in modeling.

2. The ............. button is used to identify the overlapping sketched entities.

3. The **Entity** tool is used to find out all the information about the selected entity. (T/F)

4. The .............. tool in the **Relation** panel is used to display dimensions by their designated names in place of their dimension values in the sketch.

5. The ............... tool is used to import sketches or drawing files in the sketch you are working on.

# FOR STUDENT NOTES

# FOR STUDENT NOTES

Answers to Self Assessment Questions:
1. **Inspect** 2. **Overlapping Geometry** 3. **T** 4. **Switch Dimensions** 5. **File System**

# Chapter 4

# 3D Modeling Basics

## Topics Covered

The major topics covered in this chapter are:

- *Starting Part Modeling environment*
- *Basics of Planes*
- *Datum Features*
- *Curves and Sketches*
- *Importing Sketches*
- *3D Modeling tools*

# INTRODUCTION

Till now, you have learned about creating 2D sketches. But, the purpose of creating 2D sketches is, to create 3D models that can contain the features of a real model. The tools and options to create 3D models are available in the Part mode of Creo Parametric. The method of starting Part mode is given next.

# STARTING PART MODE

This is the first step in the world of 3D Modeling using Creo Parametric. The steps to start Part mode are given next.

* Start Creo Parametric by using the icon from desktop.
* Click on the **New** button from the **Data** panel in the **Ribbon** or **Quick Access Toolbar**. The **New** dialog box will be displayed; refer to Figure-1.
* Select the **Part** radio button from the **Type** area and select the **Solid** radio button from the **Sub-type** area of the dialog box.
* Click in the **Name** edit box and specify the desired name for the file.
* Clear the **Use default template** check box because we might be working on different units in real-time.
* Click on the **OK** button from the dialog box. The **New File Options** dialog box will be displayed; refer to Figure-2.

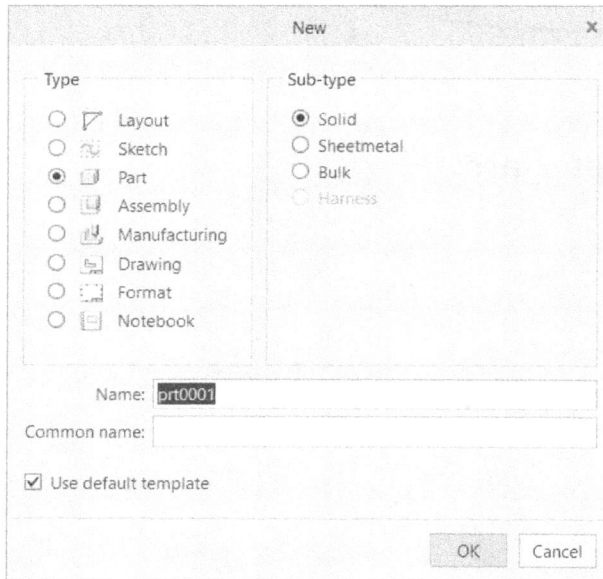

Figure-1. New dialog box

Figure-2. New File Options dialog box

* Select the desired template from the list box (in our case mmns part solid). The meaning of various templates have already been discussed in Chapter 1.
* Click on the **OK** button from the dialog box. The **Part** mode of Creo Parametric will be displayed; refer to Figure-3.

*Figure-3. Part mode of Creo Parametric*

Note that there are three default planes displayed in the Part mode of the Creo Parametric. These planes act as base for other features. Before we step into 3D Modeling, we need to understand Planes and then Datum Features.

# PLANES IN 3D

As discussed earlier, planes act as base for creating 3D models. In Creo Parametric, we have three default planes named as RIGHT, TOP, and FRONT. The selection of a plane for creating features depends on the orientation in which we want to create the model. You can easily understand the location of default planes from Figure-4.

*Figure-4. Planes*

Note that this box is oriented in the same way as the planes are oriented by default in Creo Parametric. You can select a plane by click on its name in the **Model Tree**.

## DATUMS

Datums are the references for other features. Some of the well known entities that come under datum features are:

- Planes
- Axes
- Points
- Coordinate Systems
- Curves
- Sketches

You have learned about sketches earlier in the book. Now, we will work on other datum features. All the tools for datum features are available in the **Datum** panel; refer to Figure-5.

*Figure-5. Datum panel*

The tools in this panel are discussed next.

We will first learn to create datum axis, points and Coordinate System. Later, we will work on the **Plane** tool.

## Datum Axis

The **Axis** tool is used to create an axis in the model. The axis can be used to make reference for planes or can be used to revolve the sketch for creating revolve feature. The steps to create datum axis are given next.

- Click on the **Axis** tool from the **Datum** panel. The **Datum Axis** dialog box will be displayed; refer to Figure-6 and you are asked to select entities for creating axis.

*Figure-6. Datum Axis dialog box*

- Select a plane perpendicular to which you want to create the axis. An axis perpendicular to the selected plane will be displayed; refer to Figure-7.

*Figure-7. Perpendicular axis*

- The green handles displayed are used to select the references for locating the axis with respect to the other planes.
- Drag the green handles to the plane or entity to which you want to reference the axis. You can also use the **Offset references** selection box in the dialog box to make reference selection. To do so, click in the **Offset references** selection box and select the two planes/entities while holding the CTRL key; refer to Figure-8.

*Figure-8. Axis after selecting references*

Now, we will use datum points to create axes. You will learn about the datum points in the next topic. You can go to the next topic and come back to practice the axis creation using points.

### Selecting plane and point

- If you select a plane and a point for creating the axis while holding the CTRL key, then an axis perpendicular to the plane will be created which passes through the selected point; refer to Figure-9.

*Figure-9. Axis though point and plane*

### Selecting point and axis

- If you select a point and an axis while holding the CTRL key then an axis passing through the point and normal/parallel to the selected axis will be created; refer to Figure-10. Note that if you select an edge of a solid model then an axis through the edge will be created in the same way.

*Figure-10. Axis parallel to other axis*

- To make this axis normal to the selected axis, click on the **Parallel** option next to the axis in the **References** area of the **Datum Axis** dialog box. A drop-down will be displayed.
- Click on the **Normal** option from the drop-down. The axis will become normal to the selected axis; refer to Figure-11.

*Figure-11. Axis normal to selected axis*

## Selecting 2 Points

- If you select two points while holding the CTRL key, then an axis passing through both the points will be created; refer to Figure-12.

*Figure-12. Axis through two points*

## Datum Point

The **Point** tool is used to create datum points in the model. The point can be used to make reference for planes, axes, curves, and so on. The steps to create datum point are given next.

- Click on the **Point** tool from the **Datum** panel. The **Datum Point** dialog box is displayed; refer to Figure-13 and you are asked to select entities for creating point.

*Figure-13. Datum Point dialog box*

- Click on a plane. The point will be placed and you are asked to specify references for the point; refer to Figure-14.

*Figure-14. Datum Point placed*

- Drag the handles or use the Offset references to select the references for locating the point.
- You can also select vertex of a solid model. The datum point will be created on the selected vertex.

## Plane

Plane acts as a transparent floor for any feature. All the sketches that we made in earlier chapters are made on top planes. If we bring those sketches in the part environment then you can easily find it. By default, three planes are displayed in part environment named; TOP, FRONT, and RIGHT. These planes also act as reference for the planes that we are going to create in the next steps:

- Click on the **Plane** tool from the **Datum** panel in the **Model** tab of the **Ribbon**. The **Datum Plane** dialog box will be displayed; refer to Figure-15. Also, you are asked to select the entities.

*Figure-15. Datum Plane dialog box*

## Selecting a Plane

- After selecting the **Plane** tool, click on one of the default planes available. An offset plane will be created; refer to Figure-16.
- Specify the desired offset distance in the edit box of the **Datum Plane** dialog box.
- After specifying the desired distance, click on the **OK** button from the dialog box to create the plane.

*Figure-16. Offset plane created*

## Selecting Axis and Plane

- After selecting the **Plane** tool from **Datum** panel, select an axis and a plane parallel to the selected axis while holding the **CTRL** key. The plane will be displayed as shown in Figure-17.
- Double-click on the angle value to change it.

*Figure-17. Datum plane with axis selected*

- You can also create a plane by selecting two parallel axes; refer to Figure-18.

*Figure-18. Datum plane by two axes*

- Similarly, you can create a plane by selecting an axis and a point; refer to Figure-19.

*Figure-19. Plane by axis and point*

## Selecting Points

- After selecting the **Plane** tool, one by one select the datum points while holding the control key to create a plane. The plane will be created passing through the selected points; refer to Figure-20.

*Figure-20. Datum plane created by points*

## Coordinate System

Coordinate System is also used as reference for other features. When we will work on tools like **Toroidal Bend** and **Spine Bend**, then we will need the coordinate systems to create features. The procedure to create coordinate system is given next.

*   Click on the **Coordinate System** tool from the **Datum** panel in the **Ribbon**. You are asked to specify the position of the coordinate system. Also, the **Coordinate System** dialog box will be displayed; refer to Figure-21.

Figure-21. Coordinate System dialog box

*   Click on the plane, axis, or point to specify the location of the coordinate system. If you select a plane, you are asked to specify location of the coordinate system with respect to two references. If you select an axis then you need to select one more axis as reference. If you select a point then the coordinate system will be placed on the point. To specifying the orientation of the coordinate system, you need to select edges. We will learn more about them in practical sessions.

## Curves

Curves are used to reference features that need path for their creation like, sweep, swept blend, and so on. There are three tools available to create datum curves:

1.  Curve Through Points
2.  Curve From Equation
3.  Curve From Cross Section

Figure-22. Tools for Datum Curves

## Curve Through Points

The **Curve Through Points** tool, as the name suggests, is used to create datum curve passing through the selected points. The procedure to create curves by using this tool is given next.

• Select the **Curve Through Points** tool from the **Curve** menu in the expanded **Datum** panel of the **Ribbon**; refer to Figure-22. The **Curve: Through Points** dashboard will be displayed in the **Ribbon**; refer to Figure-23.

*Figure-23. Curve Through Points dashboard*

• One by one select the datum point while holding the **CTRL** key. The datum curve will be generated; refer to Figure-24.

*Figure-24. Datum curve through points*

• You can change the curvature of the curve by selecting the two buttons highlighted in the box shown in Figure-24 after selecting each curve point.
• After selecting points, click on the **Ends Condition** tab in the **Ribbon**. The options will be displayed as shown in Figure-25.

*Figure-25. Ends Condition tab*

- Using the options, you can change the start and end point conditions of the curve; like you can make the curve tangent to plane/edge/axis at the start point or end points.
- For example, to make the curve tangent to selected axis at end point, click on the **End condition** drop-down and select the **Tangent** option; refer to Figure-26. You will be asked to select the entity to which the curve should be tangent.

*Figure-26. Tangent option for End condition*

- Select the axis/edge/plane as reference. Preview of the curve will be displayed; refer to Figure-27.

*Figure-27. Preview of tangent curve*

- Similarly, you can use the other options in the drop-down.

## Curve From Equation

The **Curve From Equation** tool, as the name suggests, is used to create datum curve by solving the equation. The procedure to create curves by using this tool is given next.

- Select the **Curve From Equation** tool from the **Curve** menu in the expanded **Datum** panel of the **Ribbon**; refer to Figure-22. The **Curve: From Equation** dashboard will be displayed in the **Ribbon**; refer to Figure-28. Also, you are asked to select a coordinate system as reference.

*Figure-28. Curve From Equation dashboard*

- Select the coordinate system from the modeling area.
- Click on the **Equation** button from the dashboard. The **Equation** dialog box will be displayed; refer to Figure-29.

*Figure-29. Equation dialog box*

- Specify the desired equation in the dialog box, for example:

$$X=t*2+t^2+\cos(t)$$
$$Y=\sin(t)$$
$$Z=50*\sin(t/2)$$

- Click on the **OK** button from the dialog box and then click on the **Done** button from the dashboard. The curve will be created; refer to Figure-30.

*Figure-30. Curve by equation*

## Curve From Cross-section

The **Curve From Cross-section** tool, as the name suggests, is used to create datum curve by using the cross-section of a solid model. Note that for using this tool, you must have a solid/surface cross-section in the drawing area like the one shown in Figure-31. (You will learn about creating cross-sections later in this book.)The procedure to use this tool is given next.

*Figure-31. Cross-section created*

• Click on the **Curve From Cross-section** tool from the **Curve** menu in the expanded **Datum** panel of the **Ribbon**. If there are more than one cross-section in the drawing area then the **CURVE** contextual tab will be displayed in the **Ribbon**; refer to Figure-32. If there is only one cross-section in the drawing area then the curve will be generated at the boundaries of cross-section.

*Figure-32. CURVE contextual tab*

• If you have two or more cross-sections to select from then click in the **Cross-section** drop-down in the **Ribbon** and select the desired cross-section. Preview of the curve will be displayed; refer to Figure-33.

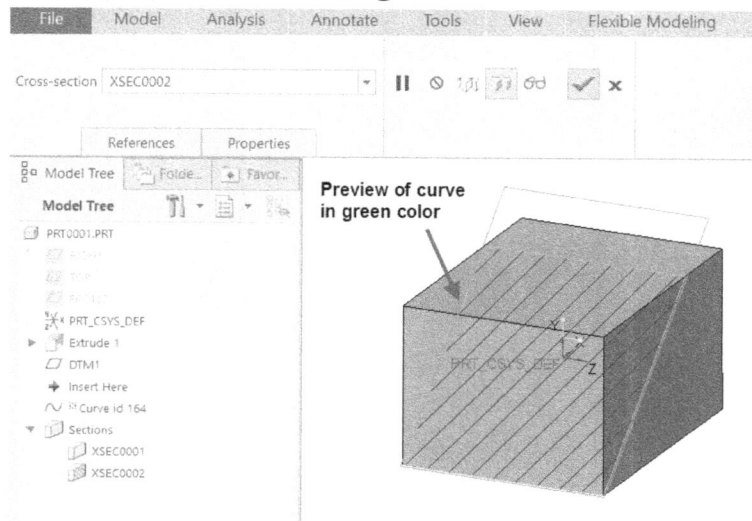

*Figure-33. Preview of cross-section curve*

- Click on the **OK** button from the **Ribbon** to create the curve.

You will learn about other tools in the **Datum** panel later in the book.

## CREATING SKETCHES ON PLANES

We have created many sketches in previous chapter. Now, we will start creating sketches on planes. The procedure to create a sketch on a plane is given next.

- Click on the **Sketch** button from the **Datum** panel in the **Ribbon**. The **Sketch** dialog box will be displayed; refer to Figure-34. Also, you are asked to select a sketching plane.

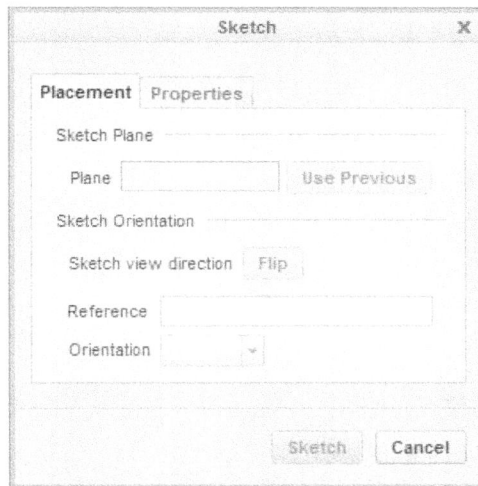

*Figure-34. Sketch dialog box*

- Select the desired plane from the planes available. The sketch will be created on the selected plane.
- On selecting a plane, the references are automatically selected in the **Reference** selection box; refer to Figure-35.

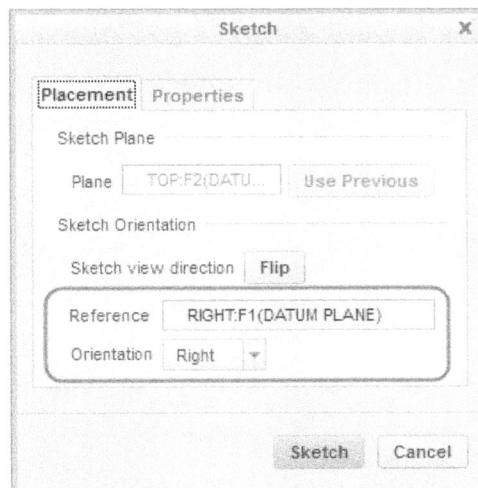

*Figure-35. Reference selection box in Sketch dialog box*

- Note that the orientation of the sketch in 3D environment is based on the selected reference plane and the options selected in the reference plane. Also, without specifying the reference plane and orientation, you can not start sketching.

- After selecting the desired planes, click on the **Sketch** button from the dialog box. The sketching plane will be displayed as shown in Figure-36. Also, the tools related to sketching will be displayed in the **Sketch** tab in the **Ribbon**.
- Click on the **Sketch View** button from the **In-Graphics** toolbar; refer to Figure-37 to make the sketching plane parallel to screen.

*Figure-36. Sketching plane*

*Figure-37. Sketch View button*

- Create the sketch by using the tools in the **Sketch** tab. The tools in the **Sketch** tab have already been discussed in previous chapters.

The sketches shown in Figure-38 show the effect of selecting different orientation for the same sketch when the sketching plane and reference planes are the same.

Orientation Right           Orientation Left

Orientation bottom           Orientation Top

*Figure-38. Effect of orientation on sketch*

After creating the sketch when you will come out of the sketching environment by clicking on the OK button, you might find that the orientation of screen is still parallel to sketching plane. To change the orientation, you need to select the Standard Orientation from the Saved Orientation drop-down in the In-Graphics toolbar; refer to Figure-39. You can also select any of the desired orientation from the drop-down like, FRONT, TOP, RIGHT, and so on.

*Figure-39. Saved Orientation drop-down*

# CREATING 3D FEATURES

Creo Parametric is a compact software with tools to create 3D features of any type. Each 3D feature tool is able to perform in many ways with combination to geometric constraints to create very simple to most complex 3D models. Tools to create 3D features are available in the **Shapes** panel of the **Model** tab in the **Ribbon**; refer to Figure-40. These tools are discussed next.

*Figure-40. Shapes panel*

# EXTRUDE

The **Extrude** tool is used to create solids by using the volume swept by a close sketch while moving along perpendicular direction. The perpendicular movement is called the height of extrusion. It is somewhat similar to extrusion process in mechanical engineering. The procedure to create extrude feature is given next.

- Click on the **Extrude** tool from the **Shapes** panel in the **Ribbon**. The **Extrude** contextual tab will be displayed; refer to Figure-41.

*Figure-41. Extrude contextual tab*

- Click on the **Placement** tab from the contextual tab. The **Placement** options will be displayed; refer to Figure-42.

*Figure-42. Placement options*

- If you have a sketch already created then select the sketch. Otherwise, click on the **Define** button from the **Placement** options. The **Sketch** dialog box will be displayed as discussed earlier.

- Select the desired planes and create the sketch. Note that sketch for extrude feature need to fulfill some requirements. You can check the requirements of the feature by clicking on the **Feature Requirements** button from the **Inspect** panel in the contextual **Sketch** tab. On selecting this button, the **Feature Requirements** dialog box will be displayed showing the feature requirements; refer to Figure-43.

*Figure-43. Feature requirements*

- So, make sure you have geometric entities in sketch. After creating geometric entities again check the feature requirement. You will find some more requirements. Fulfill them all. **To summarize, we need a sketch with geometric entities in the form of single or multiple closed loops not intersecting or overlapping each other**.
- On selecting a closed sketch, the preview of extrude feature is displayed in the modeling area; refer to Figure-44.
- Click in the **Extrude Height** edit box in the contextual tab or double-click on the **Extrude Height** dimension value in the modeling area. Specify the desired value of extrude height.
- To reverse the direction of extrusion, click on the **Flip** button in the contextual tab.

*Figure-44. Preview of extrude feature*

- Click on the **Depth** drop-down to modify the depth limiter; refer to Figure-45. You can also use the **Depth** button in **Mini toolbar** displayed on hovering cursor on the depth value in dimension.

*Figure-45. Depth drop-down*

- Select the desired limiter from the drop-down. If you have selected the **Blind** option ⬐ from the drop-down, then you need to specify the depth value for one side extrusion. If you have selected the **Both side** button ⬚, then the extrusion will be on both sides of the selected plane by specified extrusion height; refer to Figure-46. If you have selected the **Up to entity** button ⬐, then you will be asked to select a point, edge or face upto which you want to extrude the sketch.

*Figure-46. Extrusion both side*

- You can also create a thin feature by extrusion. To do so, click on the **Thicken Sketch** button ⌐ from the contextual tab. The options in the tab will be modified; refer to Figure-47.

*Figure-47. Thicken Sketch options*

- Specify the desired thickness for the section. The preview of thin feature will be displayed; refer to Figure-48.

*Figure-48. Thin feature preview*

- The flip button next to the thickness edit box is used to switch between one side, other side, or both side thickness with respect to the sketch.
- The modifications in the extrude feature that you specified by using the tools in the contextual tab can also be specified by using the options in the **Options** tab; refer to Figure-49.
- The **Remove material** button is used to subtract material from earlier created solid. This button is active only if there is a solid feature available in the modeling area. Also, the extrusion of sketch must intersect with the solid feature for removing material; refer to Figure-50.

*Figure-49. Options tab*

*Figure-50. Material removal*

- There is one more button **Extrude as surface** ⌂ . This button is used to create extruded surfaces. We will learn about this button later in the book when we will be discussing about surfaces.

Note that you can use **Sketch region** selection filter if you want to extrude a specific region of intersecting sketches; refer to Figure-51. Make sure you select the sketch region before activating the protrusion tools (extrude tool, revolve tool etc.).

*Figure-51. Extruding sketch region*

Before we move further in the modeling, we will discuss about importing the sketches so that you can practice on **Extrude** tool by using the sketches created in previous chapter.

## IMPORTING SKETCHES

- Click on the **Sketch** button from the **Datum** panel. The **Sketch** dialog box will be displayed as discussed earlier.
- Select the plane on which you want to place the imported sketch and click on the **Sketch** button from the dialog box. The sketching environment will be displayed as discussed earlier.
- Click on the **File System** button from the **Get Data** panel in the **Ribbon**; refer to Figure-52. The **Open** dialog box will be displayed; refer to Figure-53.

*Figure-52. File System button*

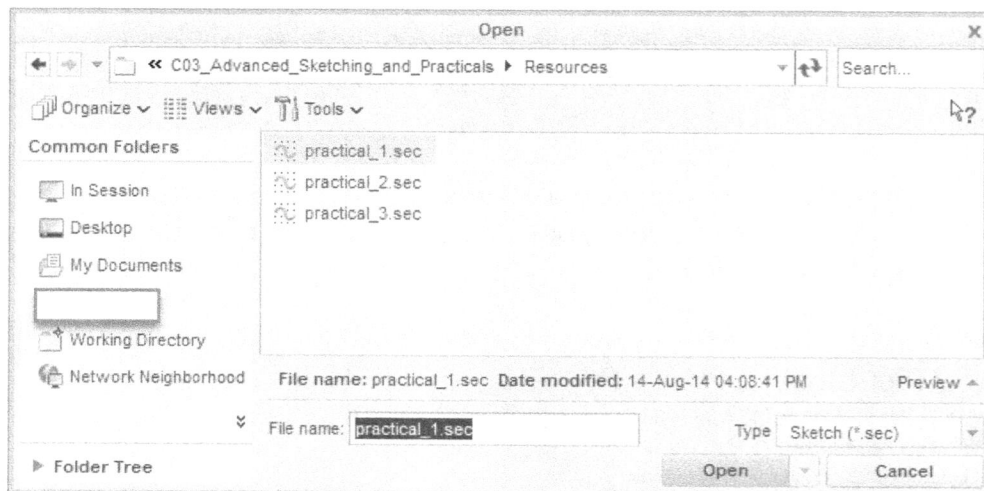

*Figure-53. Open dialog box*

- Select the desired sketch and click on the **Open** button from the dialog box. A plus sign will get attached to the cursor.
- Click at the desired place to specify the insertion point for the sketch. The **Import Section** contextual tab will be displayed along with the sketch places at the specified point; refer to Figure-54.

*Figure-54. Import Section contextual tab*

- The options in the contextual tab are similar to **Rotate Resize** tab discussed earlier. Specify the scale, rotation, and position of the sketch and then click on the **OK** button from the tab. The sketch will be placed.

# REVOLVE

The **Revolve** tool is used to create cylindrical features. You can also use the **Revolve** tool to remove material from a solid in cylindrical fashion. At most of the places, where we need cylindrical features like in modeling a pen, a shaft, and so on; we can use the **Revolve** tool to reduce the modeling time. The procedure to use this tool is given next.

- Click on the **Revolve** tool from the **Shapes** panel in the **Model** tab of **Ribbon**. The **Revolve** contextual tab will be displayed; refer to Figure-55.

*Figure-55. Revolve contextual tab*

- Click on the **Placement** tab in the contextual tab as discussed earlier and click on the **Define** button. The **Sketch** dialog box will be displayed.
- Select the desired plane from the modeling area and click on the **Sketch** button from the dialog box. The sketching environment will be displayed.
- Create a closed loop sketch with a center line not intersecting the closed loop; refer to Figure-56. Note that you need to select the **Centerline** tool from the **Datum** panel in the sketch for creating the revolve feature.

*Figure-56. Sketch for revolve feature*

- After creating the sketch, click on the **OK** button from the **Close** panel in the **Sketch** contextual tab. Preview of the revolve feature will be displayed; refer to Figure-57. Note that the gap between centerline and sketched line in the previous sketch has become hollow while revolve operation.

*Figure-57. Preview of revolve feature*

- Double-click on the dimension of 360 value displayed on the revolve feature and specify the desired value for the revolution of the sketch. Figure-58 show the same feature with revolution angle specified as 90 degree.

*Figure-58. Revolve feature with angle set as 90 degree*

- If you want to select any other centerline (axis) present in the modeling area to create the revolve feature, then click on the **Placement** tab and click on the **Internal CL** button from the tab options. You are asked to select an axis from the modeling area.
- Click on the desired axis like Z axis of coordinate system in our case; refer to Figure-59.

*Figure-59. Revolve feature by Z axis*

- Rest of the options in the contextual tab work in the same way as the options work in **Extrude** contextual tab discussed earlier.

# SWEEP

The **Sweep** tool is used to create solid/surface feature by sweeping closed loop sketch along the selected trajectory. This tool is generally used when we need to create tubes/bars following a curvature. The procedure to use this tool is given next.

- Click on the **Sweep** tool from the **Sweep** drop-down in the **Shapes** panel. The **Sweep** contextual tab will be displayed in the **Ribbon**; refer to Figure-60. Also, you will be asked to select a trajectory.

*Figure-60. Sweep contextual tab*

- Click on a sketch trajectory along which you want to sweep the closed loop; refer to Figure-61.

*Figure-61. Sketched trajectory for sweep feature*

- Click on the **Create or edit sweep section** button 🖉 to create the closed loop or section of sweep feature. The sketching environment will be activated; refer to Figure-62.

*Figure-62. Sketching environment sweep section*

- Create a close loop like circle/rectangle/polygon and then click on the **OK** button from the **Sketch** contextual tab. Preview of the **Sweep** feature will be displayed; refer to Figure-63.
- Click on the **Section Unchanged** or **Allows section to change** button as per your requirement.

*Figure-63. Preview of sweep feature*

## Sweep Feature with Multiple trajectories

- You can also create a sweep feature following more than one trajectory as shown in Figure-64. To do so, select all the trajectories one by one while holding the **CTRL** key after selecting the **Sweep** tool.

*Figure-64. Trajectories for sweep feature*

- Create a sketch which is coincident with all the trajectories; refer to Figure-65. You can use the **Coincident** button from the **Constrain** panel in the **Ribbon** to make the sketch coincident.

Section coincident
with trajectories at
points

*Figure-65. Section coincident with points*

- After creating the section, click on the **OK** button from the **Sketch** contextual tab. Preview of the sweep features will be displayed; refer to Figure-66.

*Figure-66. Sweep feature with multiple trajectory*

- Note that the rest of the options like **Thin feature** button, **Surface** button etc. are similar to **Extrude** and **Revolve**.

# HELICAL SWEEP

The **Helical Sweep** tool is used to sweep a closed loop sketch along a helical path. You can use this tool to create model of spring or you can create threads on solids by using this tool. The procedure to use this tool is given next.

- Click on the **Helical Sweep** tool from the **Sweep** drop-down in the **Shapes** panel of the **Ribbon**. The **Helical Sweep** contextual tab will be displayed in the **Ribbon**; refer to Figure-67.

*Figure-67. Helical Sweep tab in Ribbon*

- Click on the **References** tab in the **Ribbon**. The options of **References** tab will be displayed as shown in Figure-68.
- Click on the **Define** button from the tab. The **Sketch** dialog box will be displayed as discussed earlier.
- Select the desired plane (RIGHT plane in our case) and click on the **Sketch** button to start the sketching environment.
- Create sketch of path(You can use line, arc, and spline tools to create open sketch) and a centerline; refer to Figure-69.

*Figure-68. References options*

*Figure-69. Sketch for path of helical sweep*

- After creating the sketch, click on the **OK** button from the **Sketch** contextual tab. The path of helical sweep will be displayed with its pitch value; refer to Figure-70.

*Figure-70. Path of helical sweep*

- Click on the **Create or edit sweep section** button 🖉 to create profile of section for sweeping. You will enter into the sketching environment again. Note that two intersecting center lines will be displayed in the sketch denoting the starting point of the path; refer to Figure-71.
- Create the closed loop sketch of the section; refer to Figure-72. Note that the maximum vertical span of loop must be lower than the pitch, otherwise the helical sweep will not be created due to internal overlapping.

*Figure-72. Sketch for section*

*Figure-71. Starting point of path*

- Click on the **OK** button from the **Sketch** contextual tab. Preview of the helical sweep feature will be displayed; refer to Figure-73.

*Figure-73. Preview of helical sweep*

- To reverse the direction of helical sweep select one of the two buttons shown in Figure-74.

*Figure-74. Rules of sweeping*

- Change the pitch value from the **Pitch Value** edit box in the **Ribbon**.
- If you need multiple pitch for the helical sweep or say spring, then click on the **Pitch** tab in the **Ribbon**. The options will be displayed as shown in Figure-75.

*Figure-75. Pitch options*

- Click on the **Add Pitch** button from the table. A new pitch will be applied to the end point of the path. Specify the desired value for the new pitch in its relevant edit box.
- Click again on the **Add Pitch** button and you will be able to specify pitch at a desired point on the helical sweep path. Refer to Figure-76.

*Figure-76. Helical sweep with varying pitch*

- To create a tube from the solid helical sweep, click on the **Thin feature** button ⊏ and specify the desired thickness of the tube.
- Click on the **OK** button from the **Ribbon** to create the feature.

# VOLUME HELICAL SWEEP

The **Volume Helical Sweep** tool is a tool similar to helical sweep, used to remove material by revolving sketch section along helix. The result of volume Helical sweep resembles cutting a round part with rotating tool. The procedure to use this tool is given next.

- Click on the **Volume Helical Sweep** tool from the **Sweep** drop-down in **Shapes** panel of the **Ribbon**. The **Volume Helical Sweep** contextual tab will be displayed in the **Ribbon**; refer to Figure-77.

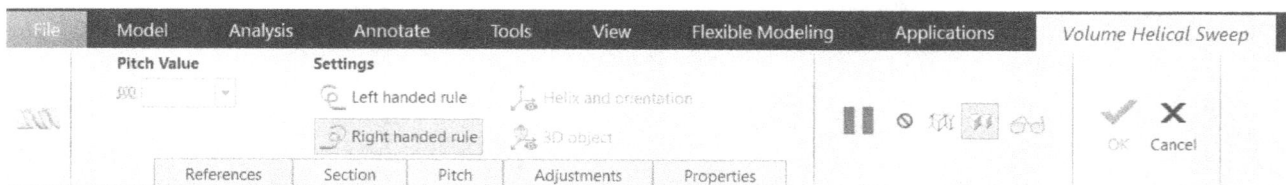

*Figure-77. Volume Helical Sweep contextual tab*

- Click on the **References** tab and then select sketch for helix profile (a line and centerline in sketch). If you have not created any sketch for profile then click on the **Define** button. The **Sketch** dialog box will be displayed asking you to select sketching plane.
- Select the desired sketching plane and draw a line & centerline in the sketch near solid body to be cut; refer to Figure-78. Click on the **OK** button from **Sketch** contextual tab to exit sketch environment.

*Figure-78. Sketch for helix profile*

- If you have created a sketch earlier for revolved section then select the **Select section** radio button and select the sketch with a close section. Click in the **Axis of revolution** selection box of **Section** tab and select the axis about which you want to revolve the section. Click in the **Origin** selection box and select a point on the close section to be used as origin; refer to Figure-79. Preview of volume helical sweep will be displayed.

*Figure-79. Selecting profile for volume helical sweep*

- If you have not created a sketch for section then select the **Sketched section** radio button from the **Section** tab and click on the **Create/Edit section** button; refer to Figure-80. The sketching environment will become active automatically.

Figure-80. Create/Edit section button

- Create the sketch of section with one line of sketch aligned to helix profile sketch earlier created; refer to Figure-81. Click on the **OK** button from **Sketch** tab after creating the sketch. Preview of volume helical sweep will be displayed; refer to Figure-82.

Figure-81. Section for volume helical sweep

Figure-82. Volume helical sweep preview

- Click in the **Pitch Value** edit box at the top-left in the **Volume Helical Sweep** contextual tab and specify the desired value of pitch.
- Select the Left handed rule or Right handed rule button from the **Ribbon** to define orientation of helical sweep.
- Select the **Helix and orientation** button to display helical path.
- Select the **Show the 3D object** button to display 3D section revolving on the part while cutting; refer to Figure-83.
- Click on the **Pitch** tab and specify the variable pitch as required. This option has been discussed in previous topic.
- Click on the **Adjustments** tab and specify the desired tilting angle for section.
- Click on the **OK** button from the **Ribbon** to create the feature.

*Figure-83. Show 3D object button*

# BLEND

As the name suggests, **Blend** tool is used to create a feature by blending two or more different sections. One of the example of such feature is a flower vase. So, we will use flower vase as an example for learning this tool. The procedure to use this tool is given next.

- Click on the **Blend** tool from the expanded **Shapes** panel; refer to Figure-84. The **Blend** contextual tab will be displayed in the **Ribbon**; refer to Figure-85.

*Figure-84. Blend tool*

*Figure-85. Blend contextual tab*

- Click on the **Sections** tab. The options for selecting and sketching the sections will be displayed; refer to Figure-86.

*Figure-86. Options in Sections tab*

- If you have a sketched section already in modeling area then click on the **Selected sections** radio button and select the sketched section. Otherwise, select the **Sketched sections** radio button and click on the **Define** button. The **Sketch** dialog box will be displayed as discussed earlier.
- Create a close loop and note down the vertices that are there in the sketch; refer to Figure-87. In this figure, there are four vertices. Also, note that there is an arrow attached to a vertex. The vertex with which the arrow is attached is the starting point of the sketch for blending.

*Figure-87. Sketch for blend*

- Click on the **OK** button from the **Sketch** contextual tab to exit. You will move back to the **Sections** options; refer to Figure-88.
- Specify the desired height value by using the blend height dimension or **Offset** edit box in the **Sections** tab.
- Click on the **Sketch** button from the **Sections** tab. The sketching environment will be displayed with a hypothetical plane selected; refer to Figure-89.

Figure-88. Blend height between two sections

Figure-89. Hypothetical plane for blend

- Create a closed loop as per your requirement. Note that the number of vertices should be the same in all the section sketches.
- In our case, we have created a circle as other section; refer to Figure-90. But since, circle do not qualify our condition of equal number of vertices, we need to use the **Divide** tool from the **Editing** panel of **Sketch** tab to make four vertices of circle.

*Figure-90. Section for blend*

- Click on the **Divide** tool and divide the circle at the points shown in Figure-91. Now, check the direction of arrow in case of circle. It is actually opposite to the arrow direction of rectangle displaying in the back ground. So, we need to change the direction of arrow.
- Select the vertex of circle with the arrow and hold the right-click. A shortcut menu will be displayed; refer to Figure-92.

*Figure-91. Division points of circle*

*Figure-92. Shortcut menu for vertex*

- Click on the **Start Point** option from the shortcut menu, the direction will get reversed; refer to Figure-93.

*Figure-93. Reversed arrow*

- Click on the **OK** button from the **Sketch** contextual tab. The preview of the blend feature will be displayed; refer to Figure-94.

*Figure-94. Preview of blend feature*

- If you want to add one more section for the blend feature, click on the **Insert** button from the **Sections** tab. A new section will be added and you will be prompted to specify the offset distance from the previous section; refer to Figure-95.

*Figure-95. Adding section for blend feature*

- Specify the desired offset distance and click on the **Sketch** button from the tab to create the section.
- Create the section with same number of vertices and then click **OK** from the **Sketch** contextual tab. Preview of the feature will be displayed; refer to Figure-96. Note that by default, the smooth blend is created but if you want to create a straight blend then click on the **Straight** radio button from the **Options** tab in the **Ribbon**; refer to Figure-97.

*Figure-96. Preview of the blend feature*

*Figure-97. Preview of blend feature with straight feature*

- If you have selected the **Smooth** radio button from the **Options** tab then the options in the **Tangency** tab are active. Using these options, you can specify the tangency conditions at the sections. For example, on selecting the **Normal** option from both start section and end section gives a blend as shown in Figure-98.

*Figure-98. Blend with normal tengency*

- Rest of the options in the **Ribbon** are same as discussed for earlier.

# ROTATIONAL BLEND

The **Rotational Blend** tool is used to create rotational blend. In the previous tool, we have specified offset distance but if we specify offset angle, then we need the **Rotational Blend** tool. The procedure to use this tool is given next.

- Click on the **Rotational Blend** tool from the expanded **Shapes** panel in the **Ribbon**. The **Rotational Blend** contextual tab will be displayed; refer to Figure-99.

*Figure-99. Rotational Blend contextual tab*

- Click on the **Sections** tab and then click on the **Define** button from the tab. The **Sketch** dialog box will be displayed as discussed earlier.
- Select the desired sketching plane and create a closed loop sketch.
- Create a datum center line at desired distance from the sketch created; refer to Figure-100.
- Click on the **OK** button from the **Sketch** contextual tab to exit the sketching environment.

- Open the **Sections** tab again and then click on the **Insert** button from the tab. The options in the tab will be displayed as shown in Figure-101.

*Figure-100. Sketch for rotational blend*

*Figure-101. Inserting sections in rotational blend*

- Specify the desired value of angle and click on the **Sketch** button.
- Create a closed loop sketch with same starting point and direction.
- Click on the **OK** button from the **Ribbon** to exit the sketching environment. Preview of the rotational blend will be displayed; refer to Figure-102.

*Figure-102. Preview of rotational blend*

- You can create more sections in the same way as we did for **Blend** tool.
- Other options in the **Ribbon** are same as we discussed earlier.

# SWEPT BLEND

The **Swept Blend** tool is used to blend two or more sections along a path. In other words, its a combination of **Sweep** tool and **Blend** tool.

- Click on the **Swept Blend** tool from the **Shapes** panel in the **Ribbon**. The **Swept Blend** contextual tab will be displayed; refer to Figure-103.

*Figure-103. Swept Blend contextual tab*

- Select a trajectories with points on trajectory equal to number of sections to be drawn. For example, in the path given in Figure-104, there are four number of points and hence we can create four sections on this path for swept blend feature.

*Figure-104. Points on path*

- Click on the **Sections** tab in the **Ribbon**. The options will be displayed as shown in Figure-105.

*Figure-105. Sections tab*

- Note that in the **Section Location** selection box, **Start** is selected. This **Start** means starting point of the selected curve. Click on the **Sketch** button from the tab to create the first section.
- Click on the **Sketch View** button from the **In-Graphics toolbar** to make the sketching plane parallel to screen. Create a sketch and note down the number of vertices in the section. Click on the **OK** button from the **Sketch** contextual tab to exit.
- Click on the **Insert** button from the tab to add a new section. Note that the End point will be selected automatically in the **Section Location** selection box.
- Click on the **Sketch** button and create a sketched section at the end point of the path; refer to Figure-106.

*Figure-106. Sections for swept blend*

- Click on the **Insert** button again and select one of the curve end point to create section; refer to Figure-107.

*Figure-107. Point to be selected*

- Click on the **Sketch** button and create a closed section with number of vertices equal to previous sections and same direction of start point as in the previous sections.

Note that if the starting points of the sections are not aligned then swept blend can be created as twisted; refer to Figure-108. After changing the start point of the middle section, the feature will be created as shown in Figure-109.

*Figure-108. Swept blend twisted due to start point locations*

*Figure-109. Swept blend feature*

- Click on the **Insert** button from the **Sections** tab and create the closed loop section on the remaining point on the curve. Preview of the swept blend feature will be displayed; refer to Figure-110.

*Figure-110. Preview of swept blend*

- Using the options in the **Tangency** tab, you can specify the tangency condition for the feature at the start and the end points.
- Click on the **Options** tab to further refine the shape of the feature and specify the desired parameters. Rest of the options in the **Swept Blend** contextual tab are same as discussed earlier.

Now, we are comfortable to create solid models with the help of solid modeling tools. But, there are some other tools that assist in creating 3D models. We will learn about them later in the book. In the next chapter, we will work on practical to create 3D models with the help of tools discussed till now.

# SELF ASSESSMENT

1. Which of the following is not default plane in Creo Parametric:

(a) RIGHT          (b) TOP
(c) FRONT          (d) LEFT

2. Which of the following are datum features in Creo Parametric:

(a) Face           (b) Surface
(c) Sketch         (d) Coordinate System

3. Which of the following is not a valid selection set for creating datum axis:
(a) a plane and a point          (b) an axis and a point
(c) two points                   (d) 2 axes

4. The .......... tool is used to create datum curve based on solution of equation.

5. In Creo Parametric, the orientation of the sketch in 3D environment is dependent on the selected reference planes. (T/F)

6. The ........... tool is used to create solids by using the volume swept by a close sketch while moving in perpendicular direction.

7. For creating solid extrude feature, we need a sketch with geometric entities in the form of single or multiple closed loops not intersecting or overlapping each other. (T/F)

8. The .......... tool is generally used when we need to create tubes/bars following a curvature.

9. The .......... tool is used to blend two or more sections along a path.

10. For create blend feature, number of corner points in the joining sketches should be same. (T/F)

**Answer to Self Assessment Questions:**

1. (d) 2. (c) and (d) 3. (d) 4. Curve From Equation 5. T 6. Extrude 7. T 8. Sweep 9. Swept Blend 10. T

# Chapter 5

# 3D Modeling
# Practical and Practice

Topics Covered

The major topics covered in this chapter are:

- **Brief note about drawing.**
- **Practical 1 to Practical 6.**
- **Practice.**

# BRIEF NOTES ABOUT DRAWINGS

Till this point, we were creating sketches and then we were creating solid features like extrude, revolve, sweep and so on with the help of those sketches. But, in real-world conditions, we are not going to have sketches. We will be finding our sketches from the engineering drawings. Being an engineer, you should be knowing about the engineering drawings but still I want to brief you on this topic.

Engineering Drawing is the exact representation of an engineering component on the paper. There are a few qualities in Engineering Drawing given as:

• It is a clear and unmistakable representation of engineering component.
• It shows all the shapes and sizes of the engineering component.
• After reading the drawing we get only one interpretation and there is no scope of confusion.

The Engineering Drawings are broadly classified into four categories:

## Machine Drawing

It is pertaining to machine parts or components. It is presented through a number of orthographic views, so that the size and shape of the component is fully understood. Part drawings and assembly drawings belong to this classification. An example of a machine drawing is given in Figure-1.

*Figure-1. Machine drawing*

# Production Drawing

A production drawing, also referred to as working drawing, should furnish all the dimensions, limits and special finishing processes such as heat treatment, honing, lapping, and surface finish to guide the craftsman on the shop floor in producing the component. The title should also mention the material used for the product, number of parts required for the assembled unit, and so on. Figure-2 shows an example of production drawing.

*Figure-2. Production drawing*

# Part Drawing

Component or part drawing is a detailed drawing of a component to facilitate its manufacture. All the principles of orthographic projection and the technique of graphic representation must be followed to communicate the details in a part drawing. A part drawing with production details is rightly called as a production drawing or working drawing.

# Assembly Drawing

A drawing that shows various parts of a machine in their correct working locations is an assembly drawing; refer to Figure-3.

*Figure-3. Assembly drawing*

**Parts List**

| Part No. | Name | Material | Qty |
|----------|------|----------|-----|
| 1 | Crank | Forged Steel | 1 |
| 2 | Crank Pin | 45C | 1 |
| 3 | Nut | MS | 1 |
| 4 | Washer | MS | 1 |

# COMPONENT REPRESENTATION METHODS

Broadly, there are two ways to present a component in engineering drawing; **Orthographic representation** and **Isometric representation**.

The orthographic representation is the method in which component is placed in the form of various views to completely define its shape and size. These orthographic views can be Front view, Right view, Top view and so on. Refer to Figure-4.

*Figure-4. Orthographic views*

The isometric representation is the method to display wire frame model of the object with its dimensions at standard angles from base planes. Note that all the objects can not be completely defined by this representation. So, it is useful to include Isometric view when we need to represent the shape of the object with some highly needed dimensions; refer to Figure-5.

*Figure-5. Isometric view*

We will be using both the representations to create our models for practice.

# PRACTICAL 1

In this practical, we will create a 3D model from the drawing given in Figure-6. The drawing has both isometric as well as orthographic views. You can use any of the two to create the model.

*Figure-6. Practical 1*

## Starting Part Modeling

- Double-click on the Creo Parametric icon from the desktop to start Creo Parametric if not started already.
- Click on the **New** button from the **Data** panel in the **Ribbon**. The **New** dialog box will be displayed.
- Specify the desired name for the file and clear the **Use default template** check box.
- Click on the **OK** button from the dialog box. The **New File Options** dialog box will be displayed as discussed earlier.
- Select the **mmns part solid** option from the dialog box and click on the **OK** button.

## Creating base sketch

- Click on the **Sketch** button from the **Datum** panel in the **Ribbon**. The **Sketch** dialog box will be displayed.
- Select the **FRONT** plane from the **Model Tree** and click on the **Sketch** button from the **Sketch** dialog box. The sketching environment will be displayed.

- Click on the **Sketch View** button from the **In-Graphics** toolbar and create the sketch as shown in Figure-7.

*Figure-7. Sketch for base*

- Click on the **OK** button from the **Sketch** contextual tab.

## Creating Extrude feature

- Click on the **Extrude** tool from the **Shapes** panel in the **Home** tab of the **Ribbon**. The **Extrude** contextual tab will be displayed.
- Select the sketch earlier created. Preview of the extrude feature will be displayed; refer to Figure-8.

*Figure-8. Preview of extrude feature*

- Specify the value of height for extrusion as **60**; refer to Figure-6 for dimension.
- Click on the **OK** button from the **Extrude** contextual tab to create the feature.

## Creating Extrude cut feature

- Click on the **Extrude** tool from the **Shapes** panel in the **Ribbon**. You are asked to select a sketching plane or sketch.

- Select the face as shown in Figure-9. The sketching environment will become active.
- Click on the **Sketch View** button from the **In-Graphics** toolbar to make the sketching plane parallel to the screen.

*Figure-9. Face to be selected for cut*

- Click on the **Project** tool from the **Sketching** panel. You are asked to select the entities to be projected on the sketching plane.
- Select the edges of the selected face as shown in Figure-10. The edges will be projected on the sketching plane; refer to Figure-11.

*Figure-10. Edges selected for projection*

*Figure-11. Edges projected*

- Using the **Line Chain** tool, create the two intersecting lines and dimension them as in Figure-12. Note that dimensions that are in red color are strong dimensions (i.e. dimension with value 15 and 30).

*Figure-12. Lines after dimensioning*

- Click on the **Delete Segment** tool from the **Editing** panel in the **Ribbon**. Drag the cursor and create a cutter curve as shown in Figure-13. The box will be created as shown in Figure-14.

Figure-13. Segment deletion

Figure-14. Box created after trimming

- Click on the **OK** button from the **Sketch** contextual tab. Preview of the extrude feature will be displayed.
- Click on the **Remove Material** button ⊿ from the **Extrude** contextual tab displayed and then click on the **Flip depth direction** button. The preview of the extrude cut feature will be displayed; refer to Figure-15.

*Figure-15. Preview of extrude cut feature*

- Click in the **Depth** value edit box of the **Extrude** contextual tab and specify the value as **20**. Preview of the feature will be displayed as shown in Figure-16.

*Figure-16. Preview of the cut feature*

- Click on the **OK** button from the **Extrude** contextual tab to create the cut feature.

## Creating Holes

- Click on the **Extrude** tool again and select the face as shown in Figure-17. The sketching environment will become active.

*Figure-17. Face to be selected for holes*

- Click on the **Sketch View** button from the **In-Graphics** toolbar to make the sketching plane parallel to the screen.
- Click on the **Center and Point** tool from the **Circle** drop-down in the **Sketching** panel of the **Sketch** contextual tab in the **Ribbon**.
- Create two circles of same radius on the selected face; refer to Figure-18. Note that dimension need not to be the same but location of circles should be similar to as shown in Figure-18.

*Figure-18. Circles created for hole*

- Dimension the circles as shown in Figure-19.

*Figure-19. Circles after dimensioning*

- Click on the **OK** button from the **Sketch** contextual tab. Preview of the extrude feature will be displayed.
- Click on the **Remove Material** button and then **Flip depth** button from the **Extrude** contextual tab.
- Click on the down arrow next to **Depth** drop-down and select the **Extrude to intersect with all surface** option.
- Click on the **OK** button from the **Extrude** contextual tab. The model will be displayed as shown in Figure-20.

*Figure-20. Model after creating holes*

# PRACTICAL 2

Create a solid model by using the drawing given in Figure-21.

(a)

(b)

*Figure-21. Practical 2*

## Starting Part Modeling

- Double-click on the Creo Parametric icon from the desktop to start Creo Parametric if not started already.
- Click on the **New** button from the **Data** panel in the **Ribbon**. The **New** dialog box will be displayed.
- Specify the desired name for the file and clear the **Use default template** check box.
- Click on the **OK** button from the dialog box. The **New File Options** dialog box will be displayed as discussed earlier.
- Select the **mmns part solid** option from the dialog box and click on the **OK** button.

## Creating base sketch

- Click on the **Sketch** button from the **Datum** panel in the **Ribbon**. The **Sketch** dialog box will be displayed.
- Select the **RIGHT** plane from the **Model Tree** and click on the **Sketch** button from the **Sketch** dialog box. The sketching environment will be displayed.
- Click on the **Sketch View** button from the **In-Graphics** toolbar and create the sketch as shown in Figure-22.

• Click on the **OK** button from the **Sketch** contextual tab to exit the sketching environment.

*Figure-22. Sketch to be created*

## Creating Base feature

• Click on the **Extrude** tool from the **Shapes** panel in the **Ribbon**.
• Select the sketch recently created, if not selected by default. Preview of the extrude feature will be displayed; refer to Figure-23.

*Figure-23. Preview of extrude*

• Click on the down arrow for **Depth** drop-down and select the **Extrude on both sides** option; refer to Figure-24.

*Figure-24. Extrude on both sides*

• Click in the **Extrude depth** edit box and specify the value as **75**. Preview of the feature will be displayed as shown in Figure-25.

*Figure-25. Extrude_feature.png*

- Click on the **OK** button from the contextual tab to create the feature.

## Creating second feature

Now, we will create the rib like structure for the model. The procedure to do so is given next.

- Click on the **Sketch** tool from the **Datum** panel in the **Ribbon** and select the **RIGHT** plane as sketching plane.
- Make sure that **TOP** plane is selected in the **Reference** selection box and **Top** is selected in the **Orientation** drop-down.
- Create the sketch as shown in Figure-26.

*Figure-26. Sketch for rib*

- Click on the **OK** button from the **Sketch** contextual tab and click on the **Extrude** tool from the **Ribbon**.
- The preview of extrude feature will be displayed. Select the **Extrude both side** option from the **Depth options** drop-down and specify the extrude depth as **12**; refer to Figure-27.
- Click on the **OK** button from the **Extrude** contextual tab to create the feature.

*Figure-27. Extruding the rib*

## Create the third feature

- Click on the **Sketch** tool from the **Datum** panel in the **Ribbon**. The **Sketch** dialog box is displayed.
- Select the top face of the second feature as sketching plane, **RIGHT** plane as reference with **Right** as orientation, and click on the **Sketch** button from the dialog box. Refer to Figure-28.

*Figure-28. Sketching face to be selected*

- Create the sketch as shown in Figure-29.

*Figure-29. Sketch for third feature*

- Click on the **OK** button from the **Sketch** contextual tab.
- Click on the **Extrude** button from the **Shapes** panel in the **Ribbon** and extrude the sketch by depth of **25**; refer to Figure-30.

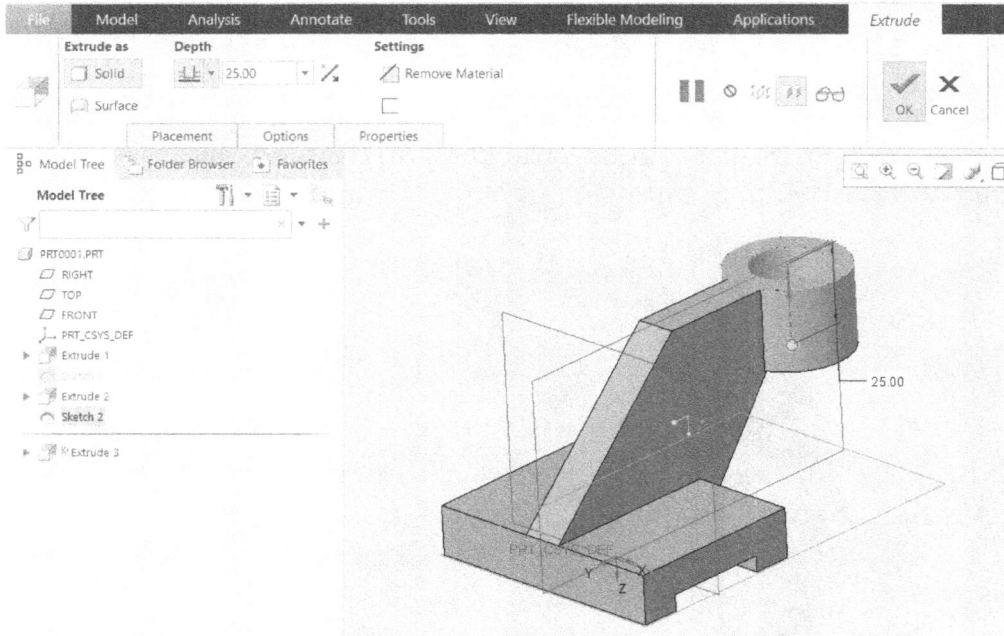

*Figure-30. Preview of the third feature*

- Click on the **OK** button from the **Extrude** contextual tab to create the feature.

# PRACTICAL 3

Create a solid model by using the drawing given in Figure-31.

*Figure-31. Practical 3*

From the above drawing, we can clearly understand that we need to create a model of a round shaft with different diameters at different locations. Here, we can use the **Revolve** tool to create the model efficiently.

## Starting Part Modeling

- Double-click on the Creo Parametric icon from the desktop to start Creo Parametric if not started already.
- Click on the **New** button from the **Data** panel in the **Ribbon**. The **New** dialog box will be displayed.
- Specify the desired name for the file and clear the **Use default template** check box.
- Click on the **OK** button from the dialog box. The **New File Options** dialog box will be displayed as discussed earlier.
- Select the **mmns part solid** option from the dialog box and click on the **OK** button.

## Creating the base sketch

- Click on the **Sketch** button from the **Datum** panel in the **Ribbon**. The **Sketch** dialog box will be displayed.
- Select the **FRONT** plane as the sketching plane and click on the **Sketch** button from the dialog box. The sketching environment will be displayed.
- Create a sketch as shown in Figure-32.

*Figure-32. Sketch for base feature*

- Make sure that you draw the centerline by using **Centerline** tool from **Datum** panel in the **Sketch** contextual tab; refer to Figure-32.
- Click on the **OK** button from the **Sketch** contextual tab to exit.

## Creating Revolve feature

- Click on the **Revolve** tool from the **Shapes** panel in the **Ribbon**. The **Revolve** contextual tab will be displayed.
- Select the sketch earlier created. Preview of the revolve feature will be displayed; refer to Figure-33.
- Make sure that **360** is specified as revolution angle and then click on the **OK** button from the **Revolve** contextual tab to create the feature.

*Figure-33. Preview of the revolve feature*

# PRACTICAL 4

In this practical, we will create a model as per the drawing given in Figure-34. The model after creation should be displayed as shown in Figure-35.

Figure-34. Practical 4

Figure-35. Model for practical 4

## Starting Part Modeling

- Double-click on the Creo Parametric icon from the desktop to start Creo Parametric if not started already.
- Click on the **New** button from the **Data** panel in the **Ribbon**. The **New** dialog box will be displayed.
- Specify the desired name for the file and clear the **Use default template** check box.

- Click on the **OK** button from the dialog box. The **New File Options** dialog box will be displayed as discussed earlier.
- Select the **mmns part solid** option from the dialog box and click on the **OK** button.

## Creating the base sketch

- Click on the **Sketch** button from the **Datum** panel in the **Ribbon**. The **Sketch** dialog box will be displayed.
- Select the **RIGHT** plane as the sketching plane and click on the **Sketch** button from the dialog box. The sketching environment will be displayed.
- Create a circle of diameter **120** mm connected with line and arc as shown in Figure-36.

*Figure-36. Sketch for path*

- Change the dimensions of the sketch as shown in Figure-37.
- Trim the unwanted portion of the path so that the path is displayed as shown in Figure-38.

*Figure-37. Sketch after dimensioning*

*Figure-38. Sketch after trimming*

- Click on the **OK** button from the **Sketch** contextual tab to exit the sketching environment.

## Creating Sweep feature

- Click on the **Sweep** tool from the **Sweep** drop-down in the **Shapes** panel of the **Ribbon**. The **Sweep** contextual tab will be displayed with sketch selected as trajectory; refer to Figure-39.

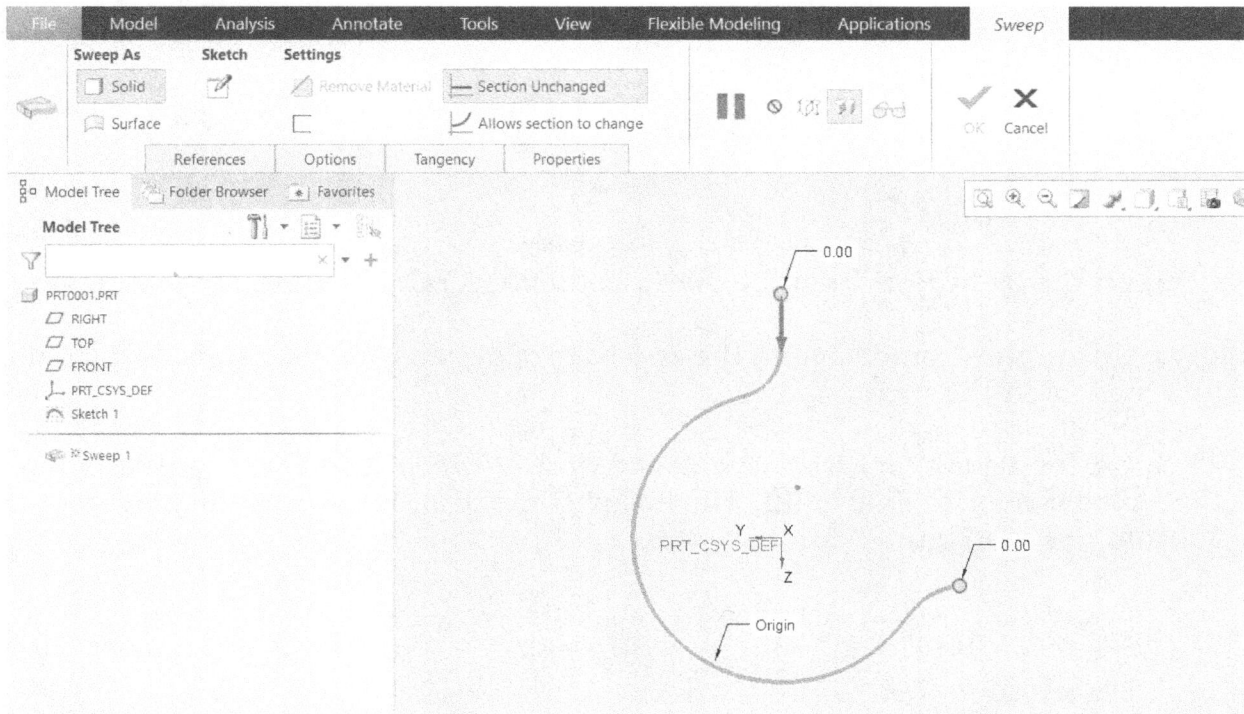

*Figure-39. Sweep contextual tab*

- Click on the **Create or edit sweep section** button 🖉 from the contextual tab. The sketching environment will be displayed.
- Click on the **Sketch View** button 🔊 from the **In-Graphics toolbar** and draw the sketch as shown in Figure-40.

*Figure-40. Sketch to create*

- Click on the **OK** button from the **Sketch** contextual tab. Preview of the sweep feature will be displayed; refer to Figure-41.

*Figure-41. Preview of sweep feature*

- Click on the **OK** button from the **Sweep** contextual tab to create the feature.

Now, we will apply a conic fillet on the end edge of sweep feature. You will learn more about this tool in the next chapter.

- Click on the **Round** tool from the **Round** drop-down in the **Engineering** panel of the **Ribbon**; refer to Figure-42. The **Round** contextual tab will be displayed in the **Ribbon**; refer to Figure-43.

*Figure-42. Round drop-down*

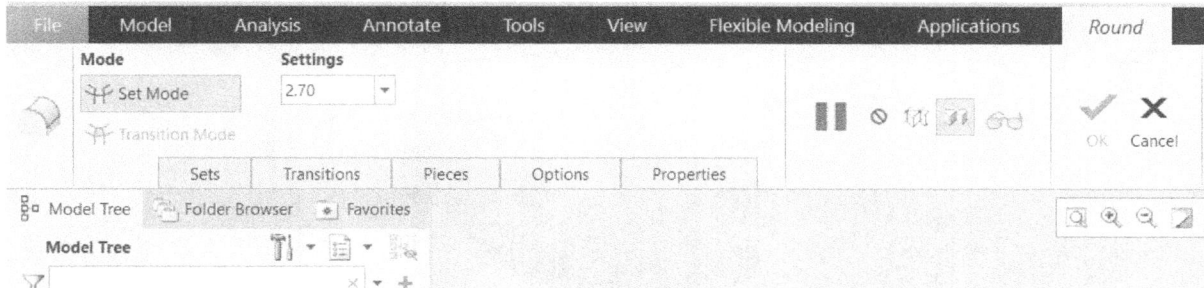

*Figure-43. Round contextual tab*

• Select the end edge of the sweep feature as shown in Figure-44. Preview of the round will be displayed; refer to Figure-45.

*Figure-44. Edge to be selected*

*Figure-45. Preview of round*

- Click on the **Sets** tab in the **Round** contextual tab. The options will be displayed as shown in Figure-46.
- Click on the **Circular** option, a drop-down will be displayed as shown in Figure-47.
- Click on the **Conic** option from the drop-down. Two values will be displayed on the preview of round; refer to Figure-48.

*Figure-46. Options in Sets tab*

Figure-47. Shape drop-down

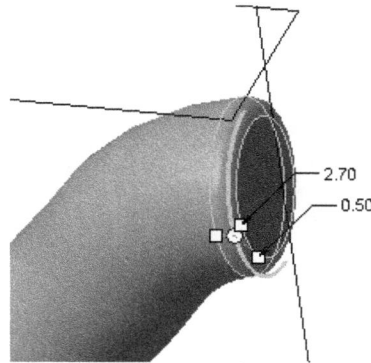

Figure-48. Preview of conic round

- Change the 0.50 value to 0.05 and change the radius i.e. 2.70 to 10. Preview of the round after changing the values will be displayed as shown in Figure-49.
- Click on the **OK** button from the **Round** contextual tab to create the feature.

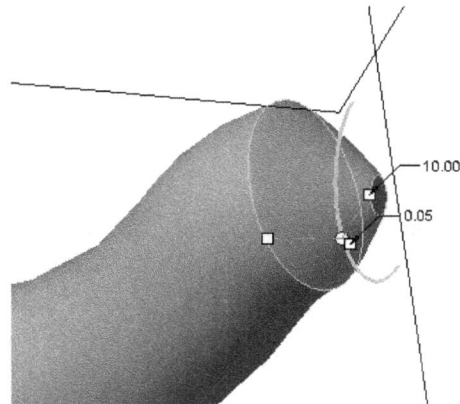

Figure-49. Preview of round after changing values

# PRACTICAL 5

In this practical, we will create a model shown in Figure-50 as per the drawing given in Figure-51. The model is of a pipe with varying section and it will require the use of datum curve.

Figure-50. Practical 5 model

*Figure-51. Practical 5*

## Starting Part Modeling

- Double-click on the Creo Parametric icon from the desktop to start Creo Parametric if not started already.
- Click on the **New** button from the **Data** panel in the **Ribbon**. The **New** dialog box will be displayed.
- Specify the desired name for the file and clear the **Use default template** check box.
- Click on the **OK** button from the dialog box. The **New File Options** dialog box will be displayed as discussed earlier.
- Select the **mmns part solid** option from the dialog box and click on the **OK** button.

## Creating the base sketches

- Click on the **Sketch** button from the **Datum** panel in the **Ribbon**. The **Sketch** dialog box will be displayed.
- Select the **FRONT** plane as the sketching plane and click on the **Sketch** button from the dialog box. The sketching environment will be displayed.
- Create datum points as shown in Figure-52.

*Figure-52. First sketch*

- Similarly, create a sketch with datum point on the TOP plane; refer to Figure-53.

*Figure-53. Second sketch*

## Creating Datum curve

- After creating the sketches, click on the **Curve through Points** tool in the **Curve** cascading menu from the expanded **Datum** panel in the **Ribbon**; refer to Figure-54. The **CURVE: Through Points** contextual tab will be displayed; refer to Figure-55.

Figure-54. Curve through Points tool

Figure-55. Curve through Points contextual tab

- Select the points one by one in the order given in Figure-56. For example, point denoted by 1 will be selected first, then 2, 3, 4 and so on.

Figure-56. Preview of curve

- To make a straight connection between the two points, select the later point from the **Placement** tab and select the **Straight line** radio button; refer to Figure-57.

*Figure-57. Straight line connection*

- Click on the **OK** button from the **CURVE: Through Points** contextual tab to create the curve.

## Creating Swept Blend feature

- Click on the **Swept Blend** tool from the **Shapes** panel in the **Ribbon**. The **Swept Blend** contextual tab will be displayed with the curve selected.
- Click on the **Sections** tab and then click on the **Insert** button from the tab.
- A new section will be added with its location at the start point of the curve; refer to Figure-58.

*Figure-58. Starting point of profile*

- Click on the **Sketch** button and create a circle of diameter **80** at the intersection point of reference planes; refer to Figure-59.
- Click on the **OK** button from the **Sketch** contextual tab.

*Figure-59. Sketch at first point*

- Similarly, create sketches on other points as per Figure-51. Preview of the Swept blend feature will be displayed as shown in Figure-60.

*Figure-60. Preview of swept blend*

- Click on the **Thin feature** button from the **Swept Blend** contextual tab. You are asked to specify the thickness value of walls in the swept blend feature; refer to Figure-61.

*Figure-61. Thickness of swept blend feature*

- Specify the thickness value as **5** in the edit box and then click on the **OK** button from the contextual tab to create the feature.

# PRACTICAL 6

In this practical, we will create model of a bolt; refer to Figure-62. Note that relations are to be applied in the dimensions to get bolts of various sizes.

*Figure-62. Practical 6*

In this practical, we will assume the value of **D** as **30** and **L** as **100**.

## Starting Part Modeling

- Double-click on the Creo Parametric icon from the desktop to start Creo Parametric if not started already.
- Click on the **New** button from the **Data** panel in the **Ribbon**. The **New** dialog box will be displayed.
- Specify the desired name for the file and clear the **Use default template** check box.
- Click on the **OK** button from the dialog box. The **New File Options** dialog box will be displayed as discussed earlier.
- Select the **mmns part solid** option from the dialog box and click on the **OK** button.

## Creating the base sketches

- Click on the **Sketch** button from the **Datum** panel in the **Ribbon**. The **Sketch** dialog box will be displayed.
- Select the **RIGHT** plane as the sketching plane and click on the **Sketch** button from the dialog box. The sketching environment will be displayed.
- Create a sketch with datum center line; refer to Figure-63.

*Figure-63. Sketch for Practical 6*

*   Click on the **OK** button from the **Sketch** contextual tab.

## Creating the Revolve Feature

*   Click on the **Revolve** tool from the **Shapes** panel in the **Ribbon**. The **Revolve** contextual tab will be displayed.
*   Select the sketch and then click on the **OK** button from the contextual tab. The revolve feature will be created; refer to Figure-64.

*Figure-64. Revolve feature created*

## Creating Extrude feature

*   Click on the **Extrude** tool from the **Shapes** panel in the **Ribbon**. The **Extrude** contextual tab will be displayed.
*   Select the top face of the model; refer to Figure-65.

*Figure-65. Face to be selected for extrude feature*

*   On doing so, the sketching environment will be displayed. Click on the **Palette** tool from **Sketching** panel in the **Ribbon**. The **Sketcher Palette** will be displayed; refer to Figure-66.

*Figure-66. Sketcher Palette dialog box*

- Double-click on the **6-Sided Hexagon** option from dialog box. Place the hexagon at the center of the circle; refer to Figure-67.

- Change the dimension of the hexagon as shown in Figure-68.

*Figure-67. Hexagon placed at center of circle*

*Figure-68. Dimensioned hexagon*

- Click on the **OK** button from the **Sketch** contextual tab. Preview of extrude feature will be displayed.
- Change the extrude depth value to **22.5** and click on the **OK** button from the **Extrude** contextual tab. The extruded bolt head will be displayed; refer to Figure-69.

*Figure-69. Extruded bolt head*

## Applying Relations

• Click on the **Annotate** tab and click on the **Show Annotations** button. The **Show Annotations** dialog box will be displayed; refer to Figure-70.

*Figure-70. Show Annotations dialog box*

• Select the model features while holding the CTRL key. The annotations will be displayed; refer to Figure-71.

*Figure-71. Model with annotations*

- Select all the check boxes under the **Show** column in the dialog box to displayed the annotations.
- Click on the **OK** button from the **Show Annotations** dialog box. Drag the dimensions out of the model for clear view.

- Click on the **Relations** tool from the expanded **Model Intent** panel in the **Model** tab of the **Ribbon**; refer to Figure-72.

*Figure-72. Relation tool*

- On selecting this tool, the **Relations** dialog box will be displayed; refer to Figure-73.

Figure-73. Relations dialog box

- Select the dimension for thickness of the bolt head. The dimension will be displayed in the **Relations** dialog box; refer to Figure-74.

Figure-74. Dimension selected for relation

- Write the formula **d7=d0*2*0.75** because thickness of bolt head is **0.75** of diameter of the bolt shank.
- Similarly, write the other formula **d8=d0*4** in the next line; refer to Figure-75.

*Figure-75. Relations dialog box with formulae*

- Click on the **OK** button from the dialog box.
- Now, change the value of shank diameter to **20** by double clicking on it while holding the **ALT** key; refer to Figure-76.

To make the changes effective, we need to regenerate the model.

*Figure-76. Diameter value to be changed*

- Click on the **Regenerate** button from the **Quick Access** toolbar; refer to Figure-77. The value of bolt head thickness and bolt head width will change automatically as per the formula.

*Figure-77. Regenerate button*

# PRACTICE 1

Create the model shown in Figure-78.

*Figure-78. Views and dimensions of the model*

# PRACTICE 2
Create the model shown in Figure-79.

*Figure-79. Views and dimensions of the model*

# PRACTICE 3

Create the model shown in Figure-80.

*Figure-80. Views and dimensions of the model*

# FOR STUDENT NOTES

# FOR STUDENT NOTES

# Chapter 6

# 3D Modeling
# Advanced

**Topics Covered**

The major topics covered in this chapter are:

- *Engineering Features like Hole, Round, Chamfer, Draft, and so.*
- *Shell, Rib, Pattern, Project, Wrap and Warp tools.*

Trajectory rib

Profile Rib

# INTRODUCTION

In the previous chapters, we have worked on the basic tools for creating 3D models. Now, we will learn about the tools that are used to add or modify the features of a base 3D model. These tools are discussed next.

# ENGINEERING FEATURES

The engineering features are the 3D model features that get their parameters from the engineering tables and act as add-on to the model. Some of the engineering features to name are: Hole, round, chamfer, rib, and so on. The tools to create these features are available in the **Engineering** panel of the **Ribbon**; refer to Figure-1. Now, we will discuss these features one by one.

*Figure-1. Engineering panel*

# Hole

The **Hole** tool is used to create holes with standard sizes or custom sizes. The standard holes can be of metric standard or unified standard. The procedure to use this tool is given next.

- Click on the **Hole** tool from the **Engineering** panel of the **Ribbon**. The **Hole** contextual tab will be displayed; refer to Figure-2.

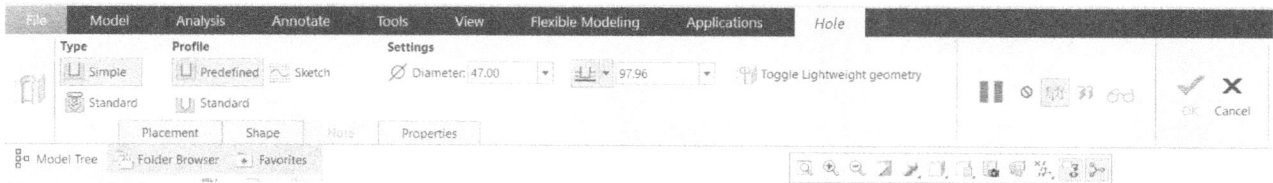

*Figure-2. Hole contextual tab*

- Click on the **Placement** tab and select the face on which you want to place the hole; refer to Figure-3.

*Figure-3. Hole placed on a face*

• Drag the two reference handles to the desired location references one by one; refer to Figure-4.

*Figure-4. References to be selected for hole*

- Change the dimensions of the hole as required. Various dimensions of hole in the model are given by Figure-5.

*Figure-5. Dimensions for hole*

- If you want to create a custom shaped hole, then click on the **Sketch profile** button in the **Hole** contextual tab. The options in contextual tab will change as shown in Figure-6.

*Figure-6. Options for sketching hole profile*

- Click on the **Open** button and select the desired profile sketch earlier created or click on the **Sketcher** button . Another application window with sketching environment will be displayed. Create the sketch of the hole profile; refer to Figure-7.

*Figure-7. Profile of the hole*

- After creating the profile, click on the **OK** button from the **Ribbon**. Preview of the hole will be displayed; refer to Figure-8.

*Figure-8. Preview of the hole*

Note that you can change the view style of the model to hidden, wire frame, no hidden, and so on; by using the options in the **Display Style** drop-down in the **In-Graphics toolbar**; refer to Figure-9.

Figure-9. Display Style drop-down

The dimensions that are shown in the previous figure are for simple holes. But in real industry, we have drills of specific sizes to create holes. So, we need to create standard holes for easy machinability.

*   To create standard holes, click on the **Standard hole** button 🔩 from the **Hole** contextual tab. The options in the dashboard will change; refer to Figure-10.

Figure-10. Hole contextual tab for standard holes

*   Select the **Add Tapping** button if you want to display the cosmetic tapping marks in the drawing. You will learn about creating drawings from model later in the book.
*   If you want to create tapered holes then click on the **Tapered** button from the contextual tab. The sizes of hole will be displayed accordingly in the hole size drop-down.
*   Select the desired standard from **ISO**, **UNC**, and **UNF** standards by using the **Standard** drop-down. If you have selected the **Add Tapping** button then specify the desired type of thread in the **Thread type** drop-down in the **Ribbon**.
*   Select the desired size from the **Size of hole** drop-down.
*   Similarly, you can change the hole to a counter-sunk or counter-bore hole. After selecting the desired button, you need specify the parameters for counter-bore/counter-sunk in the **Shape** tab of **Hole** contextual tab in **Ribbon**.
*   After setting desired parameters, click on the **OK** button from the **Ribbon**. The hole will be created and an annotation of hole will be added to the model; refer to Figure-11.

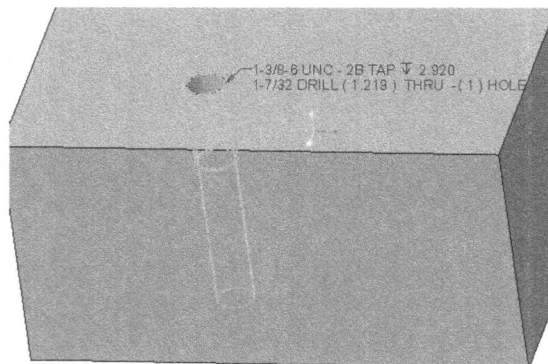

Figure-11. Standard hole with callout

## Round

The **Round** tool is used to create rounds and fillet at the sharp edges. Creating a round at the sharp edges helps in two ways:

1. It helps in reducing stress at the edges.
2. It helps in easy handling.

The procedure to create fillets by using Round tool is given next.

* Click on the **Round** tool from the **Round** drop-down; refer to Figure-12. The **Round** contextual tab will display; refer to Figure-13

*Figure-12. Round tool*

*Figure-13. Round contextual tab*

* Select an edge of the model or two intersecting faces. Preview of the round will be displayed; refer to Figure-14.

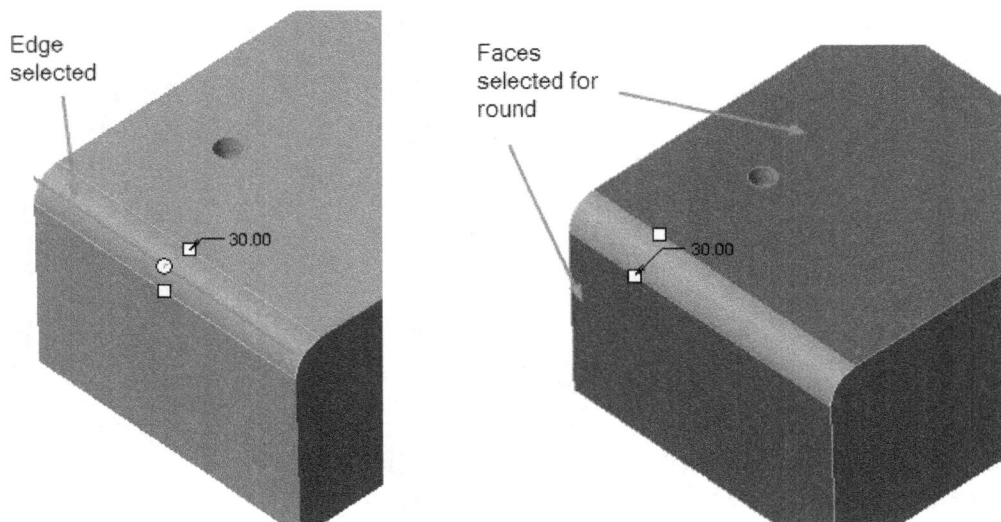

*Figure-14. Edge and faces selected for round*

- Double-click on the radius dimension in the modeling area or click in the **Radius** edit box to change the value of radius.
- Click on the **Sets** tab to change the shape of the round. The expanded **Sets** tab will be displayed as shown in Figure-15.

*Figure-15. Sets tab*

- Click on the **Circular** option in the tab. A drop-down list will be displayed; refer to Figure-16.

*Figure-16. Circular drop-down list*

- Using the options in the drop-down, you can create a round with Circular or conic shapes. We will use this option later in practical.
- If you are creating a round by using the three edges that are meeting at the same vertex as shown in Figure-17, then you can control the shape of round at the intersecting vertex by selecting the **Switch to transition mode** button.

*Figure-17. Edges for creating round*

- After clicking on the button, select the portion that is created at the intersection vertex. The options to modify the transition will be displayed in the contextual tab; refer to Figure-18.

*Figure-18. Options to change transition*

- Similarly, you can modify other options for creating round (fillet).
- After specifying the desired options, click on the **OK** button from the contextual tab to create the round feature. Note that you can select multiple faces/edges to create a fillet/fillets.

## Auto Round

The **Auto Round** tool is used to create round at the all sharp edges of the model at once. The procedure to use this tool is given next.

* Click on the **Auto Round** tool from the **Round** drop-down. The **Auto Round** contextual tab will be displayed.
* Click on the **Exclude** tab from the contextual tab and select the edges to which you do not want to apply round; refer to Figure-19.

*Figure-19. Edge not included for auto round feature*

* Specify the desired value of radius in the contextual tab and click on the **OK** button. The model will be displayed as shown in Figure-20.

*Figure-20. Model after applying auto round*

## Edge Chamfer

The **Edge Chamfer** tool is used to blunt the edges so that the stress is reduced at the edges and parts can be assembled easily. The procedure to use this tool is given next.

- Click on the **Edge Chamfer** tool from the **Chamfer** drop-down in the **Ribbon**; refer to Figure-21. The **Edge Chamfer** contextual tab will be displayed; refer to Figure-22.

*Figure-21. Chamfer tools*

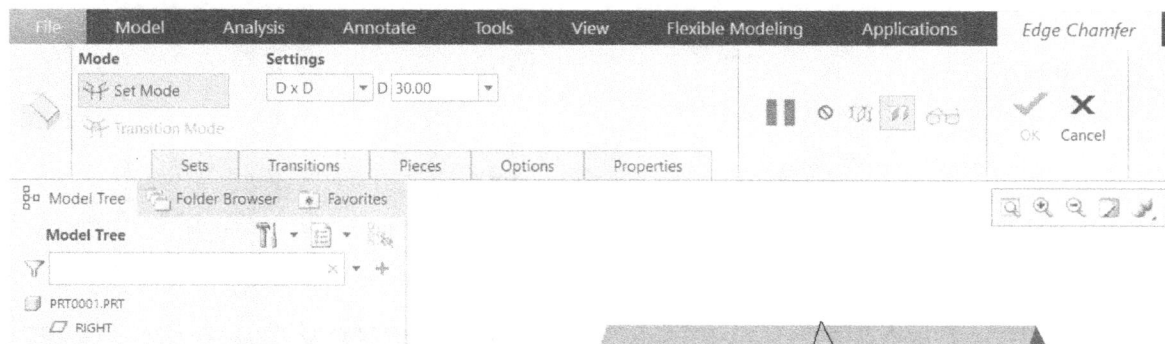

*Figure-22. Edge Chamfer contextual tab*

- Select the edges which you want to chamfer.
- Click on the **Dimension scheme** drop-down and select the desired option; refer to Figure-23.

*Figure-23. Dimension Scheme drop-down*

**Using the D1 x D2 option in the Dimension Scheme drop-down, you can independently specify the distance of chamfer edges from the chamfer centerline. Using the Angle x D option, you can specify the angle and distance value for the chamfer. Similarly, you can use the other options.**

- Specify the desired dimensions for the chamfer/chamfers and click on the **OK** button from the contextual tab. The chamfer will be created; refer to Figure-24.

*Figure-24. Chamfered model*

Note that the options like transition will work in the same way as they work for **Round** tool.

## Corner Chamfer

The **Corner Chamfer** tool is used to create chamfer at the corner created by intersection of edges. The procedure to use this tool is given next.

- Click on the **Corner Chamfer** tool from the **Chamfer** drop-down in the **Ribbon**. The **Corner Chamfer** contextual tab will be displayed; refer to Figure-25.

*Figure-25. Corner Chamfer contextual tab*

- Select a vertex, created by intersecting edges, on which you want to create corner chamfer. Preview of the chamfer will be displayed; refer to Figure-26.

*Figure-26. Preview of corner chamfer*

- Change the distance values of the chamfer as desired, by using the dimension values displayed in the modeling area.

## Draft

The **Draft** tool is used to apply taper to the faces. Draft is mostly used in mold designing, where you need to extract the molded part from the cavity. You will learn more about molding later in the book. The procedure to use this tool is given next.

- Click on the **Draft** tool from the **Draft** drop-down in the **Ribbon**. The **Draft** contextual tab will be displayed; refer to Figure-27.

*Figure-27. Draft contextual tab*

- Click on the **References** tab. The options in the tab will be displayed as shown in Figure-28.

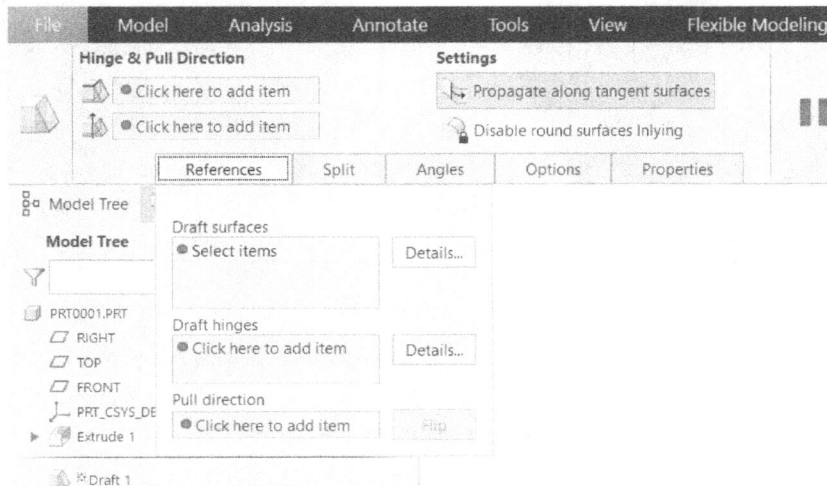

*Figure-28. References tab*

- Click in the **Draft surfaces** selection box and select the face that you want to make tapered; refer to Figure-29.

Face
selected
as draft
surface

Individual Surfaces

*Figure-29. Face selected as draft surface*

- Click in the **Draft hinges** selection box and select the edge about which you want to rotate the face for applying draft; refer to Figure-30.
- Click in the **Pull direction** selection box and select a reference (face, edge, or axis) to define the pull direction for the draft; refer to Figure-31.
- Double-click on the angle value and specify the desired draft angle. You can specify a negative value to draft in opposite direction or you can click on the **Flip** button next to **Pull direction** selection box to reverse the direction.

Chain

Edge selected as draft hinge

*Figure-30. Edge selected as draft hinge*

1.0

Face selected to
define pull direction

Pull direction
perpendicular
to the face

*Figure-31. Face selected to define pull direction*

- Creo Parametric 5.0 onwards, you can now apply draft to faces which have round or chamfer applied on their edges; refer to Figure-32.

*Figure-32. Applying draft to faces with chamfer and round*

- Select the **Propagates along tangent surfaces** button ⬚ to include nearby tangent faces automatically in draft; refer to Figure-33.

*Figure-33. Including tangent faces in draft*

- Select the **Disable round surfaces Inlying** button to exclude rounds at the transition faces; refer to Figure-34.

*Figure-34. Excluding round at transition faces*

- If you want to apply multiple angled draft to the face, then right-click on the white dot displayed on the edge selected as hinge; refer to Figure-35.
- Select the **Add Angle** option displayed in the shortcut menu; refer to Figure-36.

*Figure-35. Shortcut menu for multiple angle draft*

*Figure-36. Preview of multiple angle draft*

- You can add more angle values by right-clicking on the white dot and following the same procedure.
- Change the distance values and draft angle values as desired; refer to Figure-37.

*Figure-37. Preview of draft after applying multiple drafts*

- Click on the **OK** button from the **Draft** contextual tab to create the feature.

## Shell

The **Shell** tool is used to scoop out the material from the model. The procedure to use the **Shell** tool is given next.

- Click on the **Shell** tool from the **Engineering** panel in the **Ribbon**. The **Shell** contextual tab will be displayed; refer to Figure-38.
- Specify the desired thickness of material to be left on the walls of the model.
- Click on the **References** tab and select the face that you want to remove completely; refer to Figure-39.

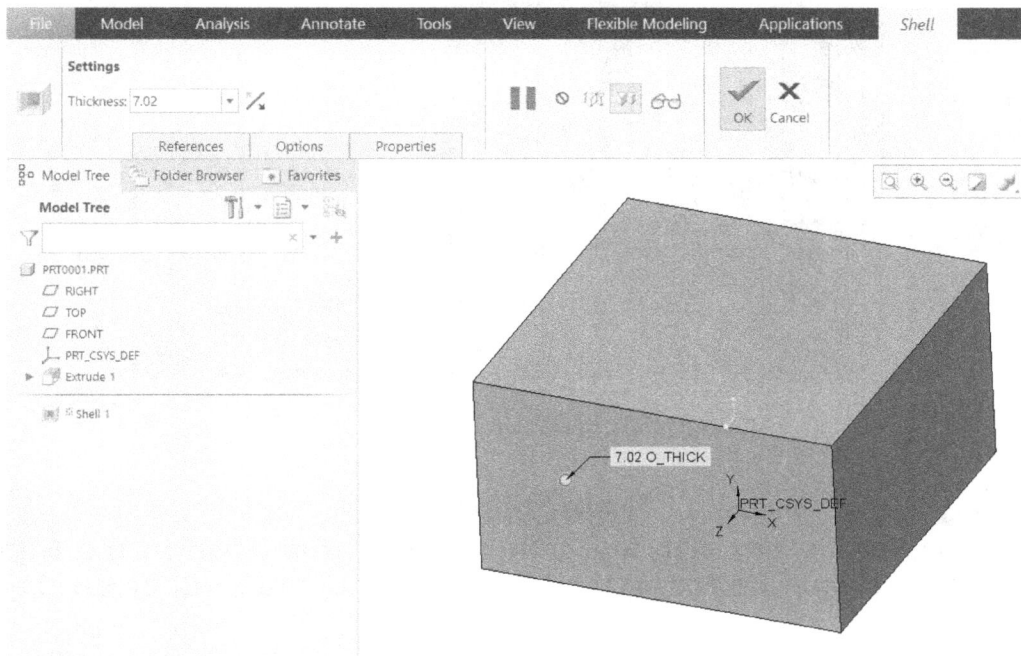

*Figure-38. Shell contextual tab*

*Figure-39. Shell feature after removing face*

# Rib

There are two type of rib features available in Creo Parametric; **Trajectory Rib** and **Profile Rib**. The Trajectory ribs are created in a location bounded by walls; refer to Figure-40. The Profile rib allows to use the selected profile while creating the rib; refer to Figure-40. The procedures to create these ribs are given next.

*Figure-40. Trajectory ribs and Profile rib*

## Trajectory Rib

- Click on the **Trajectory Rib** tool from the **Rib** drop-down in the **Engineering** panel of the **Ribbon**. The **Trajectory Rib** contextual tab will be displayed; refer to Figure-41.

*Figure-41. Trajectory Rib contextual tab*

- Click on the **Placement** tab and select the top face of the model; refer to Figure-42.

*Figure-42. Top face of the model*

- Create the sketch as shown in Figure-43.

*Figure-43. Sketch for trajectory rib*

- Click on the **OK** button from the **Sketch** contextual tab. Preview of the trajectory rib will be displayed; refer to Figure-44.

*Figure–44. Preview of trajectory rib*

- Change the thickness value and click on the **OK** button from the **Trajectory Rib** contextual tab to create the rib.

## Profile Rib

- Click on the **Profile Rib** tool from the **Rib** drop-down in the **Engineering** panel of the **Ribbon**. The **Profile Rib** contextual tab will be displayed; refer to Figure-45.

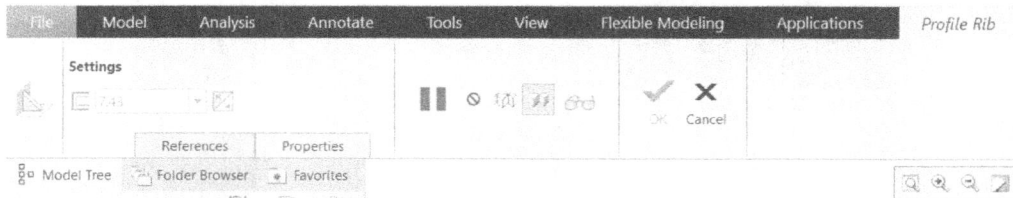

*Figure–45. Profile Rib contextual tab*

- Click on the **References** tab and click on the **Define** button from the tab; the **Sketch** dialog box will be displayed.
- Select the plane to create the profile; refer to Figure-46.

*Figure–46. Plane for creating profile*

- Create the sketch of the profile at the plane as shown in Figure-47. Note the end points of the spline created in the figure are coincident to the vertices of the model.

*Figure-47. Profile created*

- Click on the **OK** button from the **Sketch** contextual tab. Preview of the profile rib will be displayed; refer to Figure-48.

*Figure-48. Preview of profile rib*

- Specify the desired thickness value and click on the **OK** button from the **Profile Rib** contextual tab.

## Toroidal Bend

The **Toroidal Bend** tool is used to bend solids, surfaces, and datum curves to form toroidal shapes like forming tire from flat solid. The procedure to use this tool is given next.

- Click on the **Toroidal Bend** tool from the expanded **Engineering** panel in the **Ribbon**. The **Toroidal Bend** contextual tab will be displayed; refer to Figure-49.

*Figure-49. Toroidal Bend contextual tab*

- Click on the **References** tab in the **Ribbon** and then click on the **Define** button. You will be asked to select a sketching plane.
- Select the flat face of the model which is to be kept fixed during bending; refer to Figure-50. The orientation references will be selected automatically. You can change the orientation as discussed in earlier chapters.
- Click on the **Sketch** button from the dialog box and create the sketch of profile in which you want to bend the work piece; refer to Figure-51.

*Figure-50. Face selected for toroidal bend*

*Figure-51. Sketch created for bend*

- Place the datum coordinate system at the location which you want to make as neutral plane. Most of the time the coordinate system is placed on/near the sketch profile; refer to Figure-52.

Coordinate system placed at center of sketched line

*Figure-52. Placing coordinate system*

- Click on the **OK** button from the **Ribbon**. Select the **Solid Geometry** check box if you are bending a solid body from the **References** tab in the **Ribbon**. Preview of bend feature will be displayed; refer to Figure-53.

*Figure-53. Preview of toroidal bend*

- By default, the **Bend Radius** option is selected in the **Bends** drop-down in the **Ribbon**. Specify the desired radius of bend in the edit box next to the drop-down. If you want to bend the part around an axis then select the **Bend Axis** option from the drop-down and select the desired axis.
- To create a full 360 bend, select the **360 degrees Bend** option from the drop-down. You will be asked to select two faces that define the total length of solid; refer to Figure-54.

*Figure-54. Preview of 360 degree toroidal bend*

- If you want to bend a surface using the toroidal bend then after activating the tool, click on the **Quilts** selection box in the **References** tab of the **Ribbon**. You will be asked to select the surfaces. Rest of the procedure is same. In the same way, you can bend curves by using the **Curves** selection box in the **References** tab.

- Click on the **OK** button from the **Ribbon** to create the bend.

## Spinal Bend

The **Spinal Bend** tool is used to bend the selected solid body along selected curve. The procedure to use this tool is given next.

- Click on the **Spinal Bend** tool from the expanded **Engineering** panel in the **Model** tab of the **Ribbon**. The **Spinal Bend** contextual tab will be displayed; refer to Figure-55. Also, you will be asked to select a sketch entity for defining shape of solid body.

*Figure-55. Spinal Bend contextual tab*

- Select the sketch entity as shown in Figure-56. To change the origin of curve, click on the arrow displayed on the curve.

- Click in the **Bend Geometry** selection box from the contextual tab and select the solid geometry to be bent. Preview of the bend will be displayed; refer to Figure-57.

Figure-56. Sketch curve selected for spinal bend

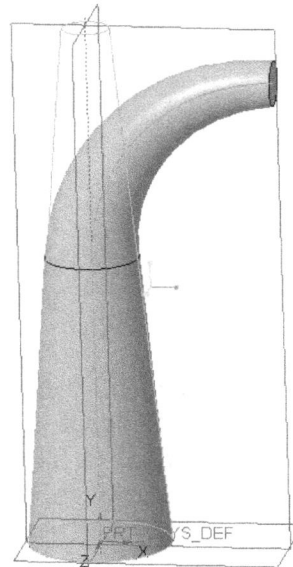

Figure-57. Preview of spinal bend

- Select the **Lock Length** check box if you want to make total length of bent solid body same as the unbent body.
- If you want to change the extent up to which you want to bend the solid body then select the **Bend Extent** drop-down in the **Ribbon**; refer to Figure-58.

Figure-58. Bend Extents drop-down

- Click on the **OK** button from the **Ribbon** to create the spinal bend feature.

## Cosmetic Sketch

The **Cosmetic Sketch** tool is used to create a sketch that is used for representation only. You can not use a cosmetic sketch as base feature for other modeling tools. The tool is available in the expanded **Engineering** panel of the **Ribbon**. The tool works in the same way as **Sketch** tool works.

## Cosmetic Thread

The **Cosmetic Thread** tool is used to a simple representation of threads. The procedure to use this tool is given next.

- Click on the **Cosmetic Thread** tool from the expanded **Engineering** panel in the **Model** tab of the **Ribbon**. The **Thread** contextual tab will be displayed in the **Ribbon**; refer to Figure-59. Also, you will be asked to select the surface on which thread representation is to be applied.

*Figure-59. Thread contextual tab*

- Select the round face of the hole/boss feature. You will be asked to select the starting face for the thread.
- Select the flat face of part from where the thread starts.
- Specify the desired pitch value in the **Pitch** edit box of **Ribbon**.
- Set the desired depth of threading in the **Depth** edit box of **Ribbon**. Preview of thread feature will be displayed; refer to Figure-60.

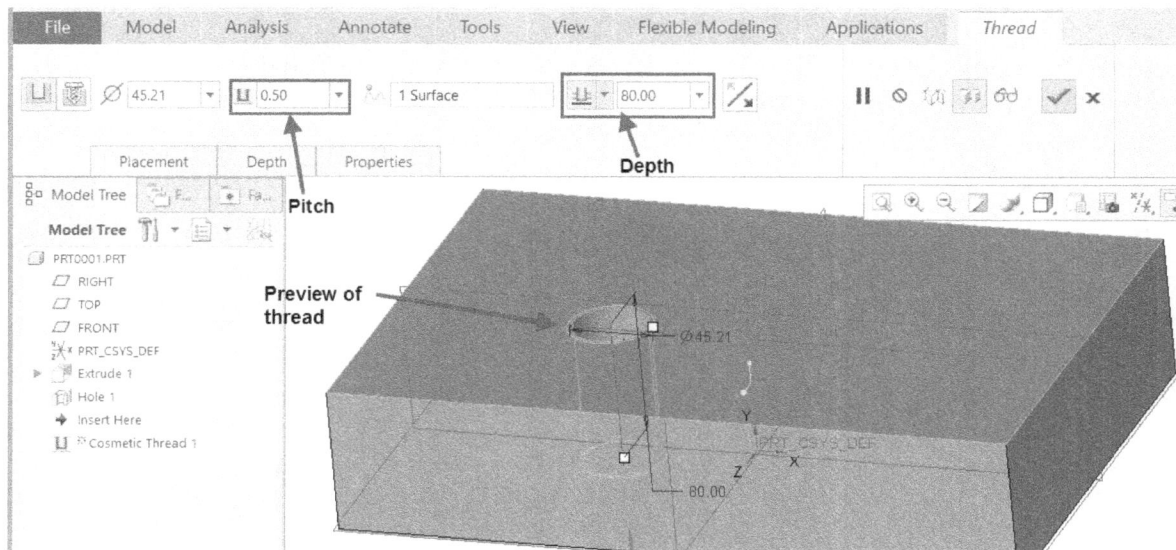

*Figure-60. Preview of Thread*

- By default, simple threads are created by this tool. If you want to create standard threads then click on the **Define Standard Thread** button from the **Ribbon**. Now, select the desired standard from the **Thread Type** drop-down and then select the desired thread size from the **Thread Size** drop-down.
- Click on the **OK** button to create the feature.

## Cosmetic Groove

The **Cosmetic Groove** tool is used to create impression of groove sketch on surface based on which groove cut can be created using Creo Manufacturing tools. The procedure to use this tool is given next.

- Click on the **Cosmetic Groove** tool from the expanded **Engineering** panel of the **Ribbon**. The **FEATURE REFS** Menu Manager will be displayed asking you to select the surface/face on which cosmetic groove is to be created.

- Select the desired surface/face and press Middle Mouse Button twice. You will be asked to select a sketching plane.
- Select the desired plane (Note that plane should be parallel to selected surface so that sketch can be projected on the surface). You will be asked to specify the sketch direction.
- Click on the **Okay** or **Flip** button as required from the Menu Manager. Set the required orientation of the sketch by using desired button from the Menu Manager.
- Create desired sketch for groove and click on the **OK** button from the **Ribbon**. The cosmetic groove will be created; refer to Figure-61.

*Figure-61. Cosmetic Groove created*

## Designated Area

The **Designated Area** tool in expanded **Engineering** panel is used to designated selected area for specific operations. The procedure to use this tool is given next.

- Create a close loop sketch on the area you want to be designated.
- Click on the **Designated Area** tool from the expanded **Engineering** panel of the **Model** tab in **Ribbon**. The **Designated Area** contextual tab will be displayed in the **Ribbon**; refer to Figure-62.

*Figure-62. Designated Area contextual tab*

- Select the closed loop sketch from model. Preview of designated area will be displayed; refer to Figure-63. You can also select the edges of model forming a closed loop for defining designated area.

*Figure-63. Preview of designated area*

• Click on the **OK** button from the **Ribbon** to create the feature.

## ECAD Area

The **ECAD Area** tool is used to assign selected area on the model to electronic circuits. The procedure to use this tool is given next.

• Click on the **ECAD Area** tool from the expanded **Engineering** panel of the **Ribbon**. The **ECAD Area** contextual tab will be displayed; refer to Figure-64.

*Figure-64. ECAD Area contextual tab*

• Click on the face of model on which you want to create ECAD area. The sketching environment will become active automatically.
• Create the boundary of electronic circuit in the form of a closed loop; refer to Figure-65. After creating sketch, click on the **OK** button from **Ribbon** to exit sketching environment.

*Figure-65. Sketch created for ECAD area*

- Click on the **Display the ecad area with hatching shown** button from the Ribbon to display the ecad area in hatching.
- Click on the **Options** tab in the **Ribbon** and select the **3D Volume** check box to create extruded volume of ecad area. Define desired depth options and depth value. Preview of ecad area will be displayed; refer to Figure-66.

*Figure-66. Preview of ECAD area*

- Set the other parameters as required and click on the **OK** button from **Ribbon** to create the feature.

You will learn about **Lattice** tool later in this book. Till this point, we have learned about the most commonly used tools in the **Engineering** panel. Now, we will discuss tools used for editing.

## EDITING TOOLS

The editing tools are available in the **Editing** panel of the **Ribbon**; refer to Figure-67. Note that very few tools are active in the **Editing** panel and rest of the tools cannot be used at present. Some of the tools like **Mirror**, **Pattern**, and so on are active after selecting a feature. Some of the tools will not be used in the solid modeling and hence will not be active.

*Figure-67. Editing panel*

The tools that are active in Solid Modeling are discussed next.

## Pattern Tools

There are three tools in the **Pattern** drop-down: **Pattern** tool, **Geometric Pattern**, and **Pattern Table**; refer to Figure-68. The **Pattern** tool is used to create multiple copies of the selected feature. The **Geometric Pattern** tool is used to create pattern of geometry. The **Pattern Table** tool is used to create a table of patterns. We will now learn about use of the **Pattern** tool and **Geometric Pattern** tool.

*Figure-68. Pattern drop-down*

## Pattern Tool

The **Pattern** tool is active only after selecting a feature of the model to be patterned. The procedure to use this tool is given next.

- Select the feature from the model that you want to pattern; refer to Figure-69.

*Figure-69. Feature selected to pattern*

- Click on the **Pattern** tool from the **Pattern Tools** drop-down. The **Pattern** contextual tab will be displayed; refer to Figure-70.

*Figure-70. Pattern contextual tab*

Now, we have different pattern types to select from the **Pattern Type** drop-down; refer to Figure-71. This drop-down is displayed on clicking on the **Dimension** option from the contextual tab.

*Figure-71. Pattern type drop-down*

We will discuss about creating a pattern by each of the option.

## Creating pattern using Dimension option

- Select the **Dimension** option from the **Pattern Type** drop-down if not selected by default. The dimensions of the selected feature will be displayed; refer to Figure-72.

*Figure-72. Dimensions displayed on model*

- Select a dimension as reference along which you want to create the pattern; refer to Figure-73.

*Figure-73. Dimension selected for first direction*

- Specify the desired distance in the active box. Preview of the pattern will be displayed; refer to Figure-74. Click on the **Dimensions** tab in the contextual tab to change the value of the distance specified.

*Figure-74. Preview of dimension pattern*

- Specify desired number of instances in the **Number of instances** edit box in the contextual tab; refer to Figure-75.

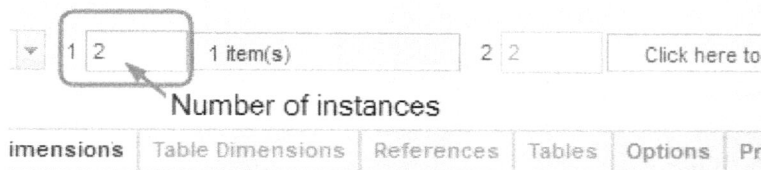

*Figure-75. Number of instances for pattern*

- Click in the **Second direction pattern dimension** collector and select the second dimension to create a rectangular pattern; refer to Figure-76.

*Figure-76. Dimension selection for second direction*

- Click on the **OK** button from the contextual tab to create the pattern.

## Creating pattern using Direction option

- Select the **Direction** option from the **Pattern Type** drop-down. The **Pattern** contextual tab will be displayed as shown in Figure-77.

*Figure-77. Pattern contextual tab for direction option*

- Select a plane/face/edge to define the first direction; refer to Figure-78.

*Figure-78. Edge selected for first direction*

- To create pattern in second direction, click in the second direction collector and select the second reference.
- Specify the desired number of instances and distance between two instances; refer to Figure-79.

*Figure-79. Edge selected for second direction*

- Click on the **OK** button from the contextual tab to create the pattern.

## Creating pattern using Direction option

- Select the **Axis** option from the **Pattern Type** drop-down. The **Pattern** contextual tab will be displayed as shown in Figure-80. Also, you are asked to select a reference.

*Figure-80. Pattern contextual tab for axis option*

- Select the axis of coordinate system or geometric centerline created. Preview of the pattern will be displayed; refer to Figure-81.

*Figure-81. Preview of axis pattern*

- Specify the number of instances in the first direction in the respective edit box in the contextual tab.
- Next, specify the angular distance between the two instances of the pattern or you can specify the total angular span; refer to Figure-82.
- Similarly, you can create angular pattern along the second direction by specifying the value in the second direction; refer to Figure-83.

*Figure-82. Options to create angular pattern*

Figure-83. Angular pattern with second direction selected

- Click on the **OK** button from the contextual tab to create the pattern; refer to Figure-84.

Figure-84. Angular pattern created

## Creating pattern using Fill option

- Select the **Fill** option from the **Pattern Type** drop-down. The **Pattern** contextual tab will be displayed as shown in Figure-85.

Figure-85. Pattern contextual tab for fill option

- Click on the **References** tab in the **Pattern** contextual tab and then click on the **Define** button; refer to Figure-86. The **Sketch** dialog box will be displayed.

*Figure-86. Define button to be selected*

- Create a closed sketch on the face of the model which you want to fill by using the instances of the selected feature to be patterned; refer to Figure-87.

*Figure-87. Sketch created for pattern*

- Specify the distance between two instances of the pattern.
- Using options of the **Grid Pattern** drop-down, you can change the grid pattern; refer to Figure-88.

*Figure-88. Grid type for fill pattern*

- After specifying the desired options, click on the **OK** button from the **Pattern** contextual tab.

## Creating pattern using Table option

- Select the **Table** option from the **Pattern Type** drop-down. The **Pattern** contextual tab will be displayed as shown in Figure-89.

*Figure-89. Pattern contextual tab for table option*

- Press the **CTRL** key and select dimensions as per which you want to create the table.
- Click on the **Edit** button, the **Pro/Table** dialog box will be displayed; refer to Figure-90.

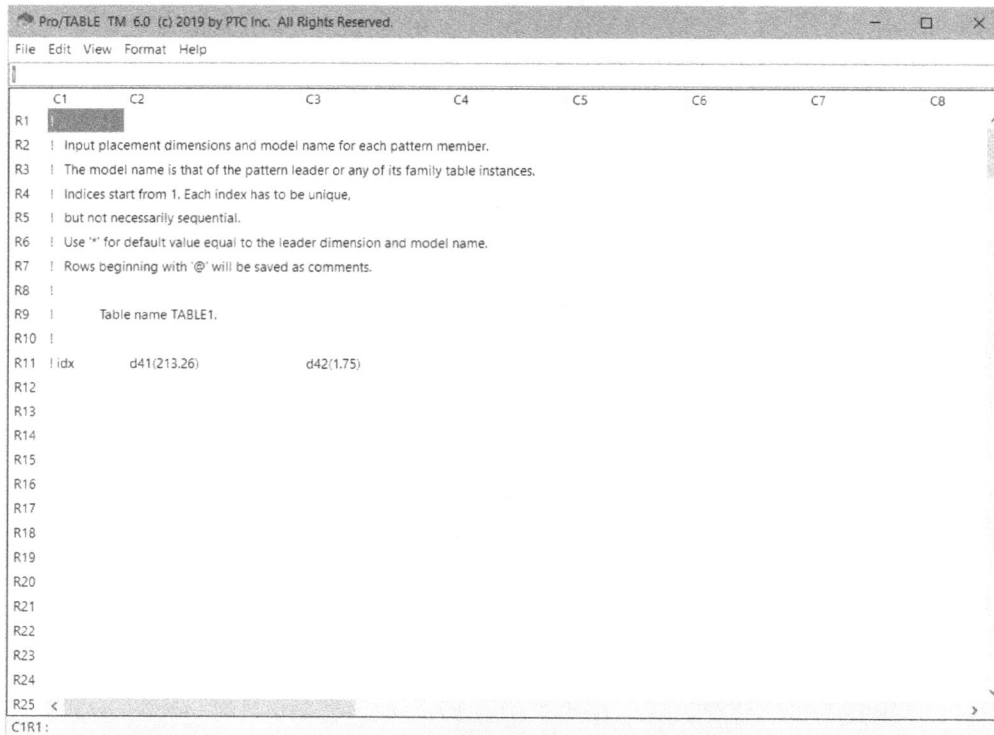

*Figure-90. Pro-Table dialog box*

- Under the **idx** column, specify the index number for pattern instance.
- Under the other columns, specify the coordinates for the holes; refer to Figure-91.

*Figure-91. Table after specifying coordinates*

- After specifying the coordinates, click on the **Exit** option from the **File** menu in the dialog box. Preview of the pattern will be displayed.
- Click on the **OK** button from the **Pattern** contextual tab.

## Creating pattern using Curve option
- Select the **Table** option from the **Pattern Type** drop-down. The **Pattern** contextual tab will be displayed as shown in Figure-92.

*Figure-92. Pattern contextual tab for curve option*

- Click on the **References** tab and click on the **Define** button from it; refer to Figure-93. The **Sketch** dialog box will be displayed.

*Figure-93. Reference tab*

- Create a curve along which you want to create the pattern; refer to Figure-94. Click on the **OK** button from the **Sketch** contextual tab. Preview of the pattern will be displayed.

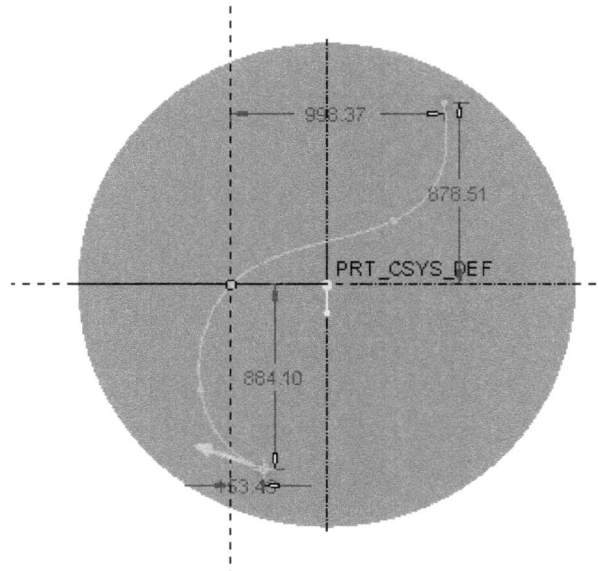

*Figure-94. Curve for pattern*

- Click on the **OK** button from the **Pattern** contextual tab.

You can use the **Point** option of the pattern in the same way as the **Curve** option do.

## Geometric Pattern

The **Geometric Pattern** tool works in the same way as the **Pattern** tool do but you can use the **Geometric Pattern** tool to also pattern the geometric entities like centerline, curves, and so on. The procedure to use this tool is given next.

- Select the entity that you want to pattern and click on the **Geometric Pattern** tool from the **Pattern** drop-down in the **Ribbon**. The **Geometric Pattern** contextual tab will be displayed as shown in Figure-95.

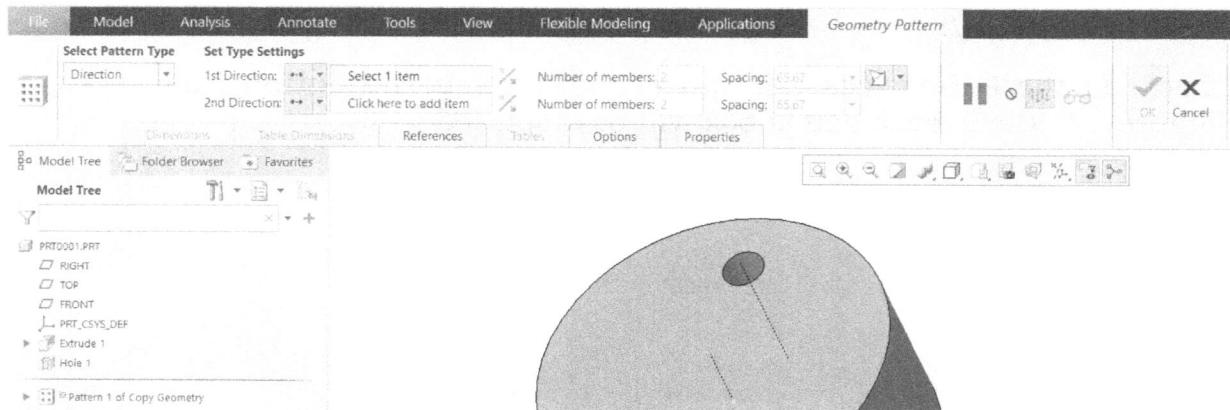

*Figure-95. Geometric Pattern contextual tab*

- The **Pattern type** drop-down has the same options as were in the Pattern contextual tab. So, we are not going to discuss these options again.
- Click on the down arrow next to **Flexible Modeling attach** button; refer to Figure-96.

*Figure-96. Options for changing pattern topology*

- The options in the drop-down are discussed next.

  **Flexible Modeling Attach** button ⊠ : Using this option, you can create geometric pattern trimmed automatically to adjust with the base surface; refer to Figure-97.

*Figure-97. Geometric pattern with Flexible modeling attach*

**Copy Geometry without attachment** button ⊟ : Using this option, you can create only geometries and the attached solid/surface is not copied; refer to Figure-98.

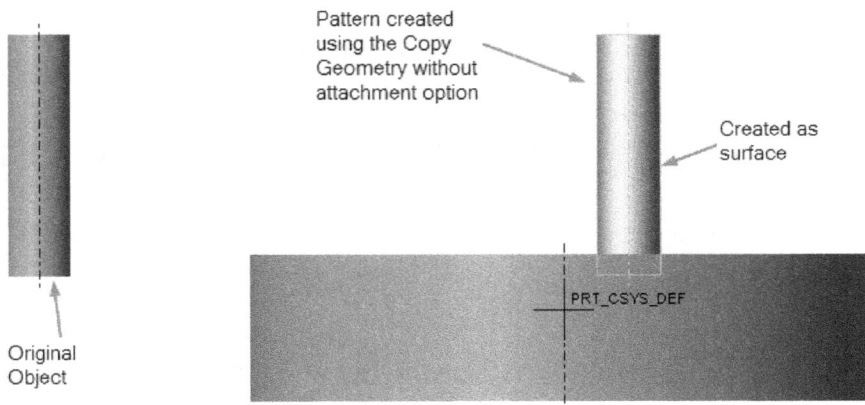

*Figure-98. Geometric pattern using copy geometry without attachment*

**Fills the volume with delimited by a quilt** button ⊟ : Using this option, you can create solid models delimited by contacting surfaces.

Similarly, you can use the other options in the drop-down.

## Pattern Table

The **Pattern Table** tool is used to manage various tables of pattern used in part or assembly. The procedure to use this tool is given next.

- Click on the **Pattern Table** tool from the **Pattern** drop-down in the **Editing** panel of the **Ribbon**. The **TABLES** dialog box will be displayed; refer to Figure-99. Note that in this example, we have already created two patterns using the **Table** option.

*Figure-99. Tables dialog box*

- Double-click on the table that you want to edit from the dialog box. The **PRO/TABLE** dialog box for selected table will be displayed; refer to Figure-100.

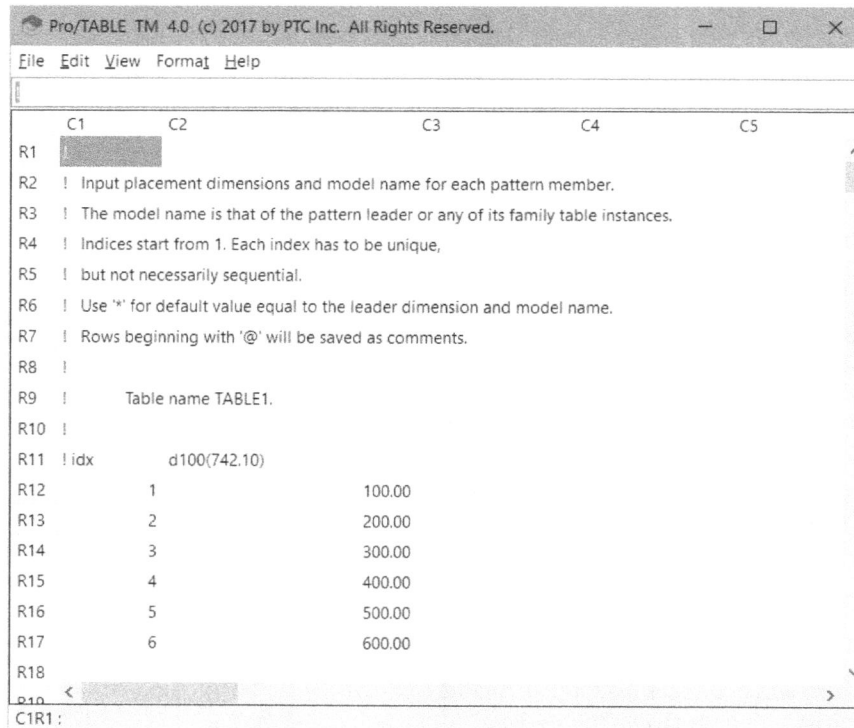

*Figure-100. Pro/Table dialog box*

- Edit the table and close the dialog box.
- To add a new table, click on the ⊞ button at the bottom in the dialog box. You will be prompted to specify name of the table. Enter the desired name of table. The **Pro/TABLE** dialog box will be displayed as discussed earlier. Specify the desired parameters and close the dialog box.
- If you want to activate the newly created table then select the table from the dialog box and click on the **Activate the selected table** ⊕ button.
- To remove a table from the list, select it and click on the **Remove** − button.
- If you want to save the selected table as external file in the Working Directory, then click on the **Write the selected table to a file** ⬚ button. Similarly, you can use the other tools of the dialog box.
- Click on the **OK** button to apply the changes and then click on the **Regenerate** button to refresh the model.

## Mirror Tool

The **Mirror** tool is used to create a mirror copy of the selected entities. The procedure to use this tool is given next.

- Select the entity/entities that you want to be mirrored.
- Click on the **Mirror** tool from the **Editing** panel in the **Ribbon**. The **Mirror** contextual tab will be displayed; refer to Figure-101.

Figure-101. Mirror contextual tab

- Select the plane about which you want to create the mirror copy; refer to Figure-102.

Figure-102. Plane selected for mirroring

- Click on the **Options** tab and clear the **Dependent Copy** check box if you want to create copy of the feature as independent entity.
- Select the **Dependent Copy** check box if you want to mirror feature to be dependent on parent feature. Select the **Partially dependent** radio button from **Options** tab to make the dimensions of mirror copy dependent on parent body. Select the Full dependent with options to vary radio button to make the mirrored body fully dependent on parent body.
- Click on the **OK** button from the **Ribbon**. The mirror copy of the entity will be created; refer to Figure-103.

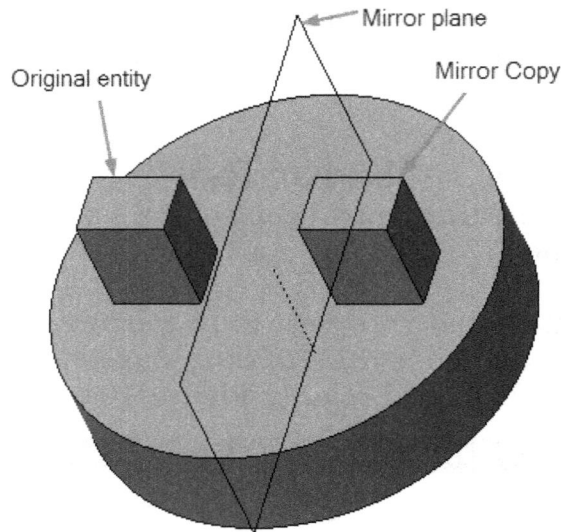

*Figure-103. Mirror copy created*

## Project Tool

The **Project** tool is used to project curves on the selected face/surface. The procedure to use this tool is given next.

- Select the **Project** tool from the **Editing** panel in the **Ribbon**. The **Projected Curve** contextual tab will be displayed; refer to Figure-104.

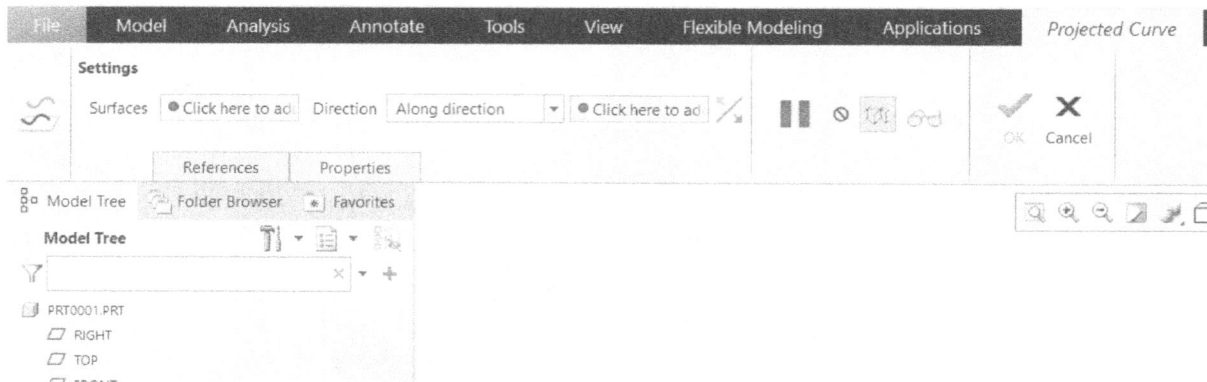

*Figure-104. Projected Curve contextual tab*

- Click on the **References** tab in the **Ribbon**. The expanded **References** tab will be displayed; refer to Figure-105.

*Figure-105. References tab for Project tool*

- Click in the **Chains** collector in the **References** tab and select the curves which you want to project.
- Click in the **Surfaces** collector and select the surface/surfaces/face/faces on which you want to project the curves.
- Click in the **Direction Reference** collector and select a plane/edge for referencing direction of projection.
- Preview of the projected curves will be displayed; refer to Figure-106.

*Figure-106. Curve projected*

- Click in the **Direction** drop-down in the **Project Curve** contextual tab and select the desired option from the drop-down; refer to Figure-107.

*Figure-107. Direction drop-down*

- Selecting the **Along direction** option generates a projected curve in the specified direction; refer to Figure-108.

*Figure-108. Curve projected using along direction option*

- Selecting the **Normal to surface** option generates a projected curve normal to the selected surface/surfaces; refer to Figure-109.

*Figure-109. Curve projected using normal to surface option*

- Click on the **OK** button from the contextual tab to created the projected curve.

## Wrap Tool

The **Wrap** tool, as the name suggests, is used to wrap a curve around the selected cylindrical surface. The procedure to use this tool is given next.

- Click on the **Wrap** tool from the expanded **Editing** panel in the **Ribbon**; refer to Figure-110. The **Wrap** contextual tab will be displayed; refer to Figure-111.

*Figure-110. Wrap tool*

*Figure-111. Wrap contextual tab*

- Click in the **References** tab and click on the **Define** button from the tab. The **Sketch** dialog box will be displayed.
- Create a sketched curve with coordinate system; refer to Figure-112. Note that the sketch curve will be wrapped around the cylinder.

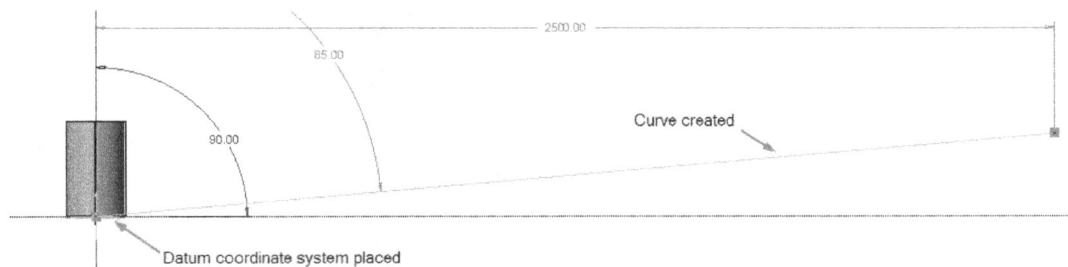

*Figure-112. Sketch created for wrap feature*

- After creating the sketch, click on the **OK** button from the **Sketch** contextual tab displayed. Preview of the wrap feature will be displayed; refer to Figure-113.

*Figure-113. Preview of wrap feature*

- If the curve you have created for wrapping is longer that the limits of selected surface then click on the **Options** tab and select the **Trim at boundary** option from it; refer to Figure-114.

*Figure-114. Trim at boundary option*

- Click on the **OK** button from the **Wrap** contextual tab to create the feature; refer to Figure-115.

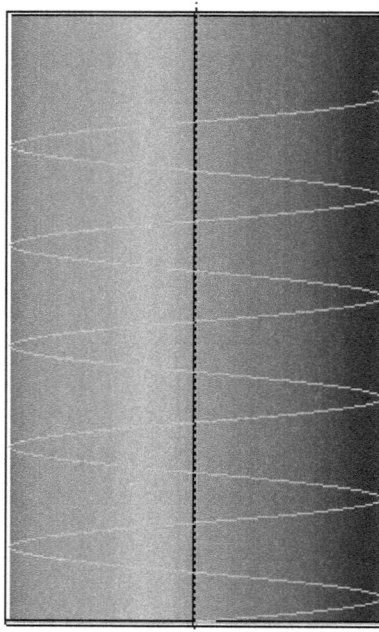

*Figure-115. Wrap feature created*

## Warp Tool

The **Warp** tool is used to modify the shape of selected solid model. The procedure to use this tool is given next.

* Click on the **Warp** tool from the expanded **Editing** panel in the **Ribbon**. The **Warp** contextual tab will be displayed; refer to Figure-116.

*Figure-116. Warp contextual tab*

* Select the solid model that you want to deform. The options in the **Warp** contextual tab will become active; refer to Figure-117.

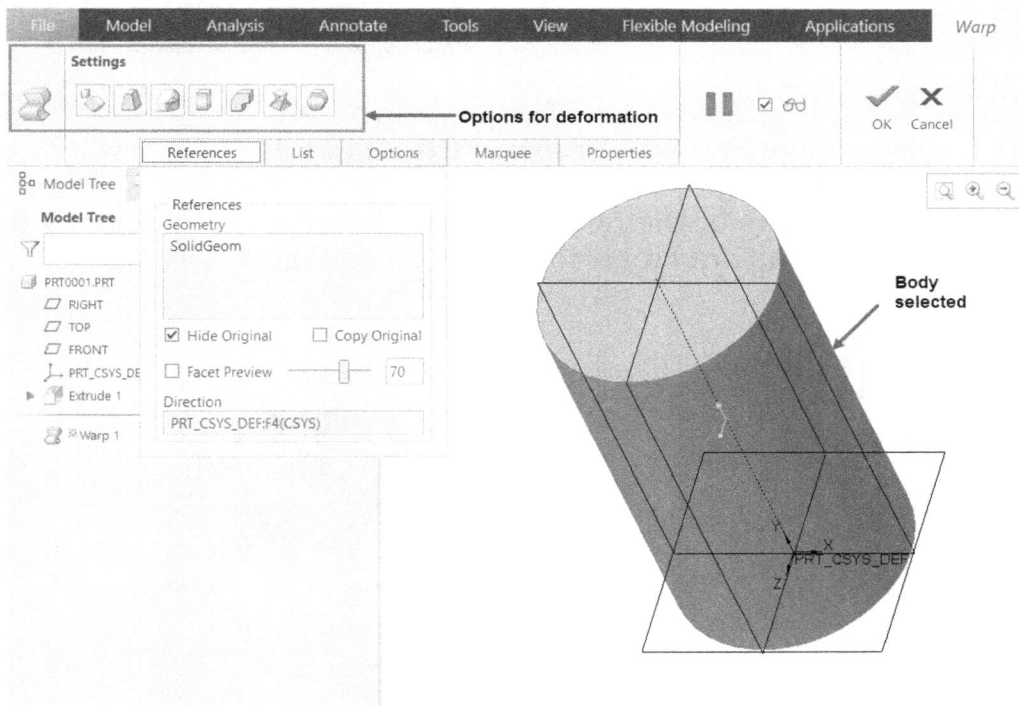

*Figure-117. Warp contextual tab with active options*

The use of each deformation option is given next.

## Using the Transform option

* Click on the **Transform** option from the **Wrap** contextual tab. The model will be displayed as shown in Figure-118.
* Move the cursor on any of the node and drag it along the directions displayed. The model will be transformed as desired; refer to Figure-119.

Nodes for Transform -ation

*Figure-118. Model after selecting Transform option*

*Figure-119. After transformation*

## Using the Warp option

- Click on the **Warp** option from the **Wrap** contextual tab. The model will be displayed as shown in Figure-120.
- Move the cursor on any of the node and drag it along the directions displayed. The model will be transformed as desired; refer to Figure-121.

*Figure-120. Model after selecting Warp option*

*Figure-121. Model after applying warp option*

## Using the Spine option

*   For using the **Spine** option of **Wrap** tool, we must have a spline already created in the modeling area; refer to Figure-122.

*Figure-122. Spline sketched*

*   Now, start the **Warp** tool, select the model and click on the **Spine** option from the **Wrap** contextual tab. You are asked to the spline. Select the spline sketched. The model will be displayed as shown in Figure-123.
*   Move the cursor on any of the node and drag it along the directions displayed. The model will be transformed as desired; refer to Figure-124.

*Figure-123. Model after selecting Spline option*

*Figure-124. Model after using Spline option*

*   Similarly, you can use the other options.

Note that for engineering point of view, the **Warp** tool does not find much significance but if you want to display deformed objects then you can use this tool.

In the next chapter, we will work on some practical and then we will practice on some complex models.

# SELF ASSESSMENT

1. The **Hole** tool is available in ................ panel of the **Ribbon**.

(a) **Shapes**          (b) **Engineering**
(c) **Datum**          (d) **Editing**

2. Which of the following tools is used to create rounds and fillet at the sharp edges.

(a) **Draft**          (b) **Shell**
(c) **Round**          (d) **Chamfer**

3. The .......... tool is used to create round at the all sharp edges of the model at once.

4. The ............. tool is used to create chamfer at the corner created by intersection of edges.

5. The **Shell** tool is used to apply taper to the faces. (T/F)

6. The **Toroidal Bend** tool can be used to form tire from flat solid. (T/F)

7. The cosmetic sketch as an important base feature for other modeling tools. (T/F)

8. The **Pattern** tool is active only after selecting a feature of the model to be patterned. (T/F)

9. The ........... tool is used to project curves on the selected face/surface.

10. Which of the following tool is used to modify shape of selected solid body using drag points?

(a) Wrap          (b) Warp
(c) Toroidal Bend  (d) Spinal Bend

**Answers to Self Assessment Questions**

# Chapter 7

# 3D Modeling Advanced Practical and Practice

## Topics Covered

The major topics covered in this chapter are:

- *Practical 1*
- *Practical 2*
- *Practice 1 to Practice 7*

# PRACTICAL 1

In this practical, we will create the model of a lever as shown in Figure-1.

*Figure-1. Practical 1*

## Starting Part Modeling

- Double-click on the Creo Parametric icon from the desktop to start Creo Parametric if not started already.
- Click on the **New** button from the **Data** panel in the **Ribbon**. The **New** dialog box will be displayed.
- Specify the desired name for the file and clear the **Use default template** check box.
- Click on the **OK** button from the dialog box. The **New File Options** dialog box will be displayed as discussed earlier.
- Select the **mmns part solid** option from the dialog box and click on the **OK** button.

## Creating Revolve feature

- Click on the **Revolve** button from the **Shapes** panel in the **Ribbon**. The **Revolve** contextual tab will be displayed.
- Select the **RIGHT** plane from the **Model Tree**. The sketching environment will be displayed.
- Click on the **Sketch View** button from the **In-Graphics** toolbar to make the sketching plane parallel to the screen if not gets by default.
- Create the sketch with geometric centerline as shown in Figure-2.

*Figure-2. Sketch for revolve*

- Click on the **OK** button from the **Sketch** contextual tab. Preview of the revolve feature will be displayed.
- Click on the **OK** button from the **Revolve** contextual tab. The feature will be displayed; refer to Figure-3.

*Figure-3. Revolve feature created*

## Creating another revolve feature

- Click on the **Revolve** tool from the **Shapes** panel in the **Ribbon**. The **Revolve** contextual tab will be displayed.
- Select the **RIGHT** plane and create the sketch as shown in Figure-4.

*Figure-4. Sketch for other revolve feature*

- Click on the **OK** button from the **Sketch** contextual tab and then from the **Revolve** contextual tab. The model will be displayed as shown in Figure-5.

*Figure-5. Model with revolved features*

## Creating the Extrude feature

- Click on the **Extrude** tool from the **Shapes** panel in the **Ribbon**. The **Extrude** contextual tab will be displayed.
- Select the **RIGHT** plane from the **Model Tree**. The sketching environment will be displayed.
- Click on the **Sketch View** button from the **In-Graphics** toolbar.
- Create a line coincident to the two vertex of the revolve features; refer to Figure-6.

*Figure-6. Line coincident to vertices*

- Click on the **Offset** tool from the **Sketching** panel in the **Ribbon** and create a line at offset distance of 3 mm; refer to Figure-7.
- Again, create a line at an offset distance of 6 mm from the recently offsetted line; refer to Figure-8.

*Figure-7. Line offsetted*

*Figure-8. Second line offsetted*

- Delete the line firstly created and make the end points of the two offsetted lines closed by a line after extending them; refer to Figure-9. Note that you need to extend the lines by using the **Line** tool in such a way that it does not cross the revolve features.

Figure-9. Lines closed after extending

- Click on the **OK** button from the **Sketch** contextual tab. Preview of the extrude feature will be displayed; refer to Figure-10.

Figure-10. Preview of the extrude feature

- Click on the **Both side extrude** option from the drop-down; refer to Figure-11.

Figure-11. Both side extrude

- Click in the edit box and specify the value as **25**. Click on the **OK** button from the **Extrude** contextual tab. The model will be displayed as shown in Figure-12.

*Figure-12. Model after creating extrude feature*

## Creating the Rib feature

- Click on the **Profile Rib** tool from the **Rib** drop-down in the **Engineering** panel of the **Ribbon**. The **Profile Rib** contextual tab will be displayed in the **Ribbon**.
- Click on the **RIGHT** plane from the **Model Tree**. The sketching environment will be displayed. Click on the **Sketch View** button from the **In-Graphics** toolbar.
- Click on the **Line Chain** tool from the **Line** drop-down in the **Sketching** panel in the **Ribbon**. Create a line coincident with the vertices of revolve features created earlier; refer to Figure-13.

*Figure-13. Line created coincident with vertices*

- Click on the **OK** button from the **Sketch** contextual tab. Preview of the rib feature will be displayed; refer to Figure-14.

*Figure-14. Preview of Rib feature*

- Change the thickness value to **10** and then click on the **OK** button from the **Profile Rib** contextual tab.

## Creating the angular hole

- Click on the **Plane** button from the **Datum** panel in the **Ribbon**. The **Datum Plane** dialog box will be displayed.
- Select the top face of the revolve feature at the higher level; refer to Figure-15. Preview of the offset datum plane will be displayed.

Face to be selected

*Figure-15. Face to be selected for creating plane*

- Create the plane at a distance of 20 below the selected face; refer to Figure-16.

*Figure-16. Plane to be created*

- Make sure the recently created plane is selected and then click on the **Sketch** button from the **Datum** panel in the **Ribbon**. The sketching environment will become active.
- Create a datum centerline by using the **Centerline** tool from the **Datum** panel in the **Ribbon**; refer to Figure-17.

*Figure-17. Centerline created for hole*

- Click on the **OK** button from the **Close** panel in the **Sketch** contextual tab.
- Click on the **Hole** tool from the **Engineering** panel in the **Ribbon**. The **Hole** contextual tab will be displayed.
- Select the centerline recently created and select the round face of the respective revolve feature; refer to Figure-18. Preview of the hole will be displayed; refer to Figure-19.

Centerline to be selected

Face to be selected

*Figure-18. Entities to be selected for creating hole*

*Figure-19. Preview of the hole feature*

- Click on the **Use standard hole** button ⊔ from the **Hole** contextual tab and then click on the **Adds countersink** button ⋈.
- Click on the **Shape** tab and specify the values as shown in Figure-20.
- Click on the **OK** button from the contextual tab to create the hole. The hole will be created as shown in Figure-21.

*Figure-20. Values in Shape tab*

*Figure-21. Hole created*

## Creating handle of lever

- Click on the **Extrude** tool from the **Shapes** panel in the **Ribbon**. The **Extrude** contextual tab will be displayed.
- Select the face as shown in Figure-22. The sketching environment will be displayed.
- Create the sketch as shown in Figure-23.

Face to be selected

*Figure-22. Face to be selected for creating extrude feature*

60.00

25.00

Sketch created

*Figure-23. Sketch for extrude*

- Click on the **OK** button from the **Sketch** contextual tab and specify the depth of extrusion as **12**.
- Click on the **OK** button from the **Extrude** contextual tab to create the feature.

# PRACTICAL 2

Create the model as shown in Figure-24. The dimensions are given in Figure-25.

*Figure-24. Practical 2*

*Figure-25. Practical 2 drawing*

## Starting Part Modeling

- Double-click on the Creo Parametric icon from the desktop to start Creo Parametric if not started already.
- Click on the **New** button from the **Data** panel in the **Ribbon**. The **New** dialog box will be displayed.
- Specify the desired name for the file and clear the **Use default template** check box.
- Click on the **OK** button from the dialog box. The **New File Options** dialog box will be displayed as discussed earlier.
- Select the **mmns part solid** option from the dialog box and click on the **OK** button.

## Creating the base feature

- Click on the **Extrude** button from the **Shapes** panel in the **Ribbon**. The **Extrude** contextual tab will be displayed.
- Select the **TOP** plane from the **Model Tree**. The sketching environment will be displayed.
- Create the sketch as shown in Figure-26.

*Figure-26. Sketch for base feature*

- Click on the **OK** button from the **Sketch** contextual tab. Preview of the extrude feature will be displayed.
- Specify the depth value as **42** and select the both side extrude option; refer to Figure-27.

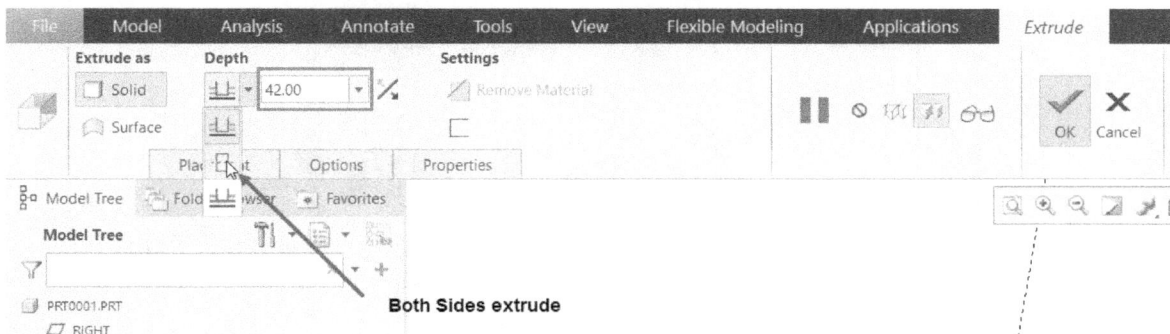

*Figure-27. Depth value for extrusion*

- Click on the **OK** button from the **Extrude** contextual tab.

## Creating the second extrude feature

- Click on the **Extrude** button from the **Shapes** panel. The **Extrude** contextual tab will display.
- Click on the top face of the base feature; refer to Figure-28. The sketching environment will get activated.

Face to be selected

*Figure-28. Top face to be selected*

- Create the sketch as shown in Figure-29.

*Figure-29. Sketch to be created*

- After creating sketch, click on the **OK** button from the **Ribbon** and extrude the feature by **7** mm downward; refer to Figure-30.

*Figure-30. Extruding second feature*

## Creating third extrude feature

- Again, click on the **Extrude** tool from the **Ribbon** and create the sketch as shown in Figure-31.

*Figure-31. Sketch for third extrusion*

- Extrude the sketch to a height of **14**; refer to Figure-32.

*Figure-32. Preview of third extrusion*

## Creating Mirror feature

• Select the second and third extruded feature and then click on the **Mirror** tool from the **Editing** panel in the **Ribbon**. The **Mirror** contextual tab will be displayed and you are asked to select a plane.

• Select the top plane as mirror reference; refer to Figure-33.

*Figure-33. Plane selected for mirror reference*

• Click on the **OK** button from the contextual tab. The mirror feature will be created; refer to Figure-34.

*Figure-34. Mirror feature created*

Till this point, we have added material in the model. Rest of the features are for removing the material.

## Creating First Extruded cut

• Select the Front plane and then click on the **Extrude** tool from the **Ribbon**. The sketching environment will be activated. Click on the **Sketch View** button from the **In-Graphics** toolbar.

• Create the sketch as shown in Figure-35.

*Figure-35. Sketch for first cut feature*

- Click on the **OK** button from the **Sketch** contextual tab. Preview of the extrude feature will be displayed.
- Click on the **Remove Material** button from the **Extrude** contextual tab.
- Select the **Both Side extrude** button from the drop-down and specify the value of extrusion; refer to Figure-36.

*Figure-36. Options specified for extrude cut*

- Click on the **OK** button from the contextual tab to create the cut feature.

## Creating Second Extruded cut

- Select the top face of the model as shown in Figure-37 and then click on the **Extrude** tool from the **Ribbon**. The sketching environment will be activated. Click on the **Sketch View** button from the **In-Graphics** toolbar.

*Figure-37. Face selected for second cut feature*

- Create the sketch as shown in Figure-38 by using the **Project** tool.

*Figure-38. Sketch for second cut*

- Click on the **OK** button and create the extrude cut up to the surface of first cut feature; refer to Figure-39.

*Figure-39. Depth of extrude cut*

## Mirroring the cut feature

- Select the second extrude cut feature and click on the **Mirror** tool from the **Ribbon**. You are asked to select the mirroring plane.

- Select the Top plane from the **Model Tree** and click on the **OK** button from the contextual tab. The mirror feature will be created; refer to Figure-40.

*Figure-40. Mirror feature of extruded cut*

## Creating Third Extruded Cut

- Click on the **Plane** tool from the **Datum** panel in the **Ribbon** and select the Top plane.
- Create a plane at an offset distance of 2.5 from the Top plane; refer to Figure-41.

*Figure-41. New plane created*

- Select the newly created plane and click on the **Extrude** tool from the **Ribbon**. The sketching environment will be displayed.
- Create the sketch as shown in Figure-42.

*Figure-42. Sketch for third cut feature*

- Click on the **OK** button from the **Sketch** contextual tab and click on the **Remove Material** button from the **Extrude** contextual tab. Preview of the extruded cut will be displayed.
- Click on the **Extrude to intersect with all surfaces** button from the drop-down in the **Extrude** contextual tab; refer to Figure-43 and then click on the **OK** button from the **Extrude** contextual tab to create the feature. Make sure that the direction of extrusion is upward.

*Figure-43. Extrude to intersect all button*

- Create mirror feature of this extruded cut with respect to the Top plane; refer to Figure-44.

*Figure-44. Mirror feature of third extruded cut*

## Creating Fillet

- Click on the **Round** tool from the **Engineering** panel in the **Ribbon**. The **Round** contextual tab will be displayed.
- Specify the value of fillet radius as **78** in the edit box available in the contextual tab.
- One by one, select the four edges as shown in Figure-45. Preview of the fillet will be displayed; refer to Figure-46.

*Figure-45. Edges selected for fillet*

*Figure-46. Preview of fillet*

• Click on the **OK** button from the contextual tab to create the feature.

# PRACTICE 1

Create the model(isometric view) as shown in Figure-47. The dimensions of the model are given in Figure-48.

*Figure-47. Model for Practice 1*

*Figure-48. Practice 1 drawing views*

# PRACTICE 2

Create the model using the drawings shown in Figure-49.

*Figure-49. Rope Pulley*

# PRACTICE 3

Create the model as shown in Figure-50. Dimensions are given in Figure-51. Assume the missing dimensions.

*Figure-50. Practice 3 model*

*Figure-51. Practice 3 Drawing*

# PRACTICE 4

Create the model by using the dimensions given in Figure-52.

*Figure-52. Practice 4*

# PRACTICE 5

Create the model by using the dimensions given in Figure-53.

*Figure-53. Practice 5*

# PRACTICE 6

Create the model by using the dimensions given in Figure-54.

*Figure-54. Practice 6*

FOR STUDENT NOTES

# Chapter 8

# 3D Modeling Advanced-II

## Topics Covered

The major topics covered in this chapter are:

- *Datum Reference*
- *Datum Graph*
- *User-Defined Feature (UDF)*
- *Scaling Model*
- *Setting Features Read-Only*
- *Family Table*
- *Importing Part/Assembly*
- *Merge/Inheritance*
- *Collapse feature*
- *Rendering*

# INTRODUCTION

In previous chapters, you have learned about 3D modeling tools and you have created many parts. The tools that are discussed till now are sufficient for common work in industries. But, you can also customize the software as per your company needs so that you can be more productive in lesser time. Along with customizing, there are some other part modeling tools which are discussed in this chapter.

# DATUM REFERENCE

Datum reference features are user-defined surface sets, edge chains, datum planes, datum axes, datum points, or datum coordinate systems. Each datum reference feature can create an intent object, which you can use to show your design intent in a model. The **Datum Reference** tool is used to create Datum reference feature. The procedure to use this tool is given next.

- Click on the **Reference** tool from the expanded **Datum** panel in the **Model** tab of the **Ribbon**. The **Datum Reference** dialog box will be displayed; refer to Figure-1.

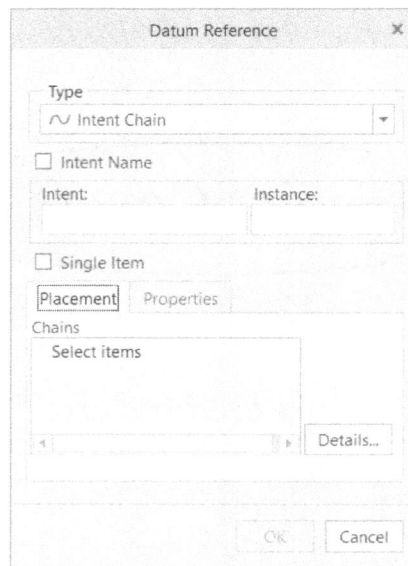

*Figure-1. Datum Reference dialog box*

- Select the type of datum reference from the **Type** drop-down in the dialog box. For example, select the **Intent Surface** option from the **Type** drop-down if you want to create surface set.
- Select the **Intent Name** check box and specify the desired name in the edit box below it to identify the selected entity set. Like, surfaces selected belong to boss feature in the mold then specify the name of the feature BOSS Feature.
- Select the entities from the model based on the type selected in the **Type** drop-down; refer to Figure-2.

*Figure-2. Surfaces selected for Intent Surface feature*

• Click on the **OK** button from the dialog box to create the feature.

Note that by default, the names of intent objects are not displayed in the **Model Tree**. To display the name of intent objects, click on the **Tree Columns** option from the **Settings** drop-down in the **Model Tree**; refer to Figure-3. The **Model Tree Columns** dialog box will be displayed; refer to Figure-4.

*Figure-3. Tree Columns option*

*Figure-4. Model Tree Colums dialog box*

Make sure the **Info** option is selected in the **Type** drop-down. Select the **Intent Name** option from the list and click on the **Add Column** >> button. Click on the **OK** button from the dialog box. A new column of intent name will be added to the **Model Tree**; refer to Figure-5.

*Figure-5. Intent Name column in Model Tree*

Now, you will ask what is the benefit of using this feature. Once you have created a datum reference feature, you can annotate your design intent, search for intent objects with the **Search** tool, define saved queries, and automatically place user-defined features (UDFs). You will learn about all these options later in this book.

## DATUM GRAPH

The datum graph is used to create a sketch graph which can be used in various relations and formulae in modeling. The procedure to create datum graph is given next.

- Click on the **Graph** tool from the expanded **Datum** panel in the **Model** tab of the **Ribbon**. You will be prompted to specify the name of the graph.
- Enter the desired name in the input box displayed. The sketching environment will be activated. Create the desired graph with coordinate system; refer to Figure-6.

*Figure-6. Graph created*

- After creating graph, click on the **OK** button from the **Ribbon**. The graph will be added in the **Model Tree**; refer to Figure-7.

*Figure-7. Graph added in Model Tree*

The main use of graph is in making relations in the model. Like, you can make the length of a block change automatically based on specified value of width using graph values; refer to Figure-8.

*Figure-8. Relation applied using graph*

In the above example, if you change the value of **d1** then using the graph value of **d2** will change automatically. Like, if **d1** is **10** then **d2** will be **15** and if **d1** is **30** then **d2** will be **20**.

# USER-DEFINED FEATURE

A User-Defined Feature is a combination of features that can be applied to part at one place. A User-Defined Feature (UDF) consists of selected features (holes/cuts etc.), all their associated dimensions, any relations between the selected features, and a list of references for placing the UDF on a part. The procedures to create and apply User-defined features are discussed next.

## Creating User-Defined Feature

- Create the model with features that you want to include in User Defined Feature (UDF); refer to Figure-10.
- Click on the **UDF Library** tool from the **Utilities** panel of the **Tools** tab in the **Ribbon**. The **UDF Menu Manager** will be displayed; refer to Figure-11.

If **Tools** tab is not displayed by default in the **Ribbon** then right-click on any tab of **Ribbon** and select the **Tools** check box in **Tabs** cascading menu of the shortcut menu; refer to Figure-9.

*Figure-9. Activating Tools tab in Ribbon*

Features created
to use as UDF

*Figure-11. UDF Menu Manager*

*Figure-10. Features for UDF*

- Click on the **Create** option from the **Menu Manager**. You will be prompted to specify the name of the UDF.
- Enter the desired name like we have specified the name as "Support". **UDF OPTIONS** sub-menu will be displayed asking you to select **Stand Alone** or **Subordinate** option from the menu; refer to Figure-12. A standalone UDF copies all the original model information into the UDF file. Because of this, a standalone UDF requires more storage space than a subordinate UDF. If you make any changes to the reference model, they are not reflected in the UDF. A subordinate UDF gets its values directly from the original model at run time, so the original model must be present for the subordinate UDF to function. If you make any changes to the dimension values in the original model, they are automatically reflected in the UDF. We will discuss the use of **Stand Alone** option. You can apply the same method for Subordinate UDF.

*Figure-12. UDF OPTIONS sub-menu*

- Select the **Stand Alone** option from sub-menu and click on the **Done** button. You will be asked whether to include the reference part or not. Click **Yes** if you want to display a preview of UDF while inserting in the Part. Click **No** if you do not want to display the preview of the UDF. ( We have selected **Yes** button). You will be asked to select the features to be included in UDF.
- Select the features from **Model Tree** while holding the CTRL key and then press Middle Mouse Button(MMB) thrice. You will be prompted to specify names of various references used for making these features.
- Enter the desired names of references (highlighted in the model) in the input box. After specifying all the reference names, press MMB.
- You can click on the **OK** button and create the UDF but the dimensional variation will not be available while inserting the feature. If you want to change the shape and size of the UDF while inserting in the part, select the **Var Dims** option from the **UDF** dialog box and click on the **Define** button; refer to Figure-13. You will be prompted to select the dimensions that you want to be changed; refer to Figure-14.

*Figure-13. UDF dialog box*

*Figure-14. Dimensions displayed for UDF*

- Select the dimensions that you want to be variable while holding the CTRL key. The selected dimensions will be highlighted in green color.
- Press the MMB thrice. You will be asked to specify names of various dimension variables.
- Enter the desired names like hole diameters; refer to Figure-15.

*Figure-15. Specifying names for dimensional variables*

- After specifying all the dimensions, click on the **OK** button from the dialog box. The **UDF Menu Manager** will be activated again.
- Click on the **Done/Return** option from the **Menu Manager**. The UDF will be saved in the current working directory.

## Inserting User-Defined Feature

- Click on the **User-Defined Feature** tool from the **Get Data** panel in the **Model** tab of the **Ribbon**. The **Open** dialog box will be displayed with UDF files in working directory; refer to Figure-16.

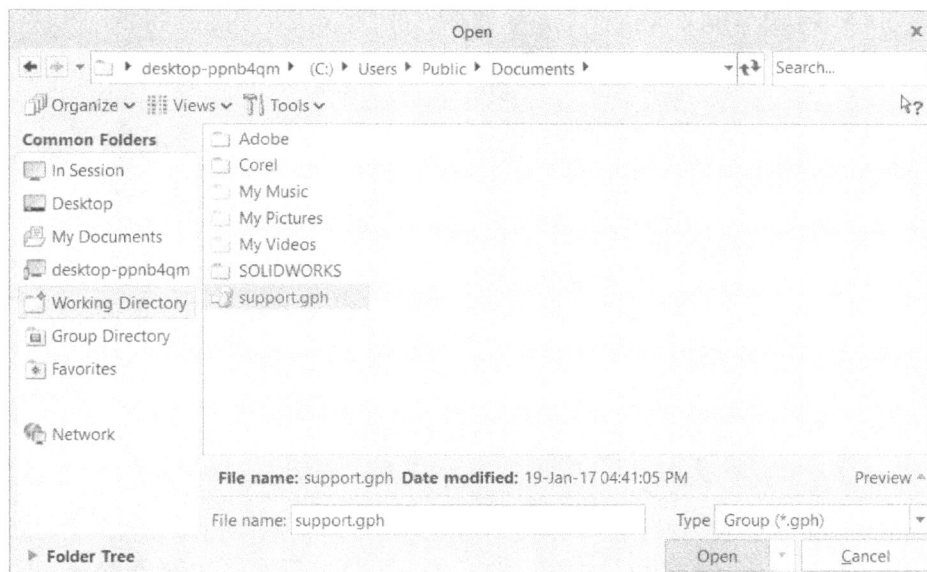

*Figure-16. Open dialog box with UDF file*

- Double-click on the desired UDF file from the dialog box. Note that the file extension for UDF file is ".gph". The **Insert User-Defined Feature** dialog box will be displayed; refer to Figure-17.

*Figure-17. Insert User-Defined Feature dialog box*

- Select the **Advanced reference configuration** check box if you want to map the UDF references to the selected references. (This option is required for placing UDF properly otherwise you might need to redefine the whole feature again in the model.) Select the **View source model** check box if you want to view the source model while inserting UDF. Select the **Make features dependent on dimensions of UDF** check box to make all the features of UDF dependent on the dimensions specified for UDF. Select the desired check box and click on the **OK** button (We have selected the **Advanced reference configuration** check box and the **View source model** check box). The **User-Defined Feature Placement** dialog box will be displayed with preview of source model; refer to Figure-18. Also, you will be asked to select the references for UDF one by one.

*Figure-18. User Defined Feature Placement dialog box with source model preview*

- Select the references based on hints specified while creating the UDF; refer to Figure-19.

*Figure-19. Reference selection for UDF*

- If UDF has variables to change then click on the **Variables** tab in the dialog box and specify the desired values; refer to Figure-20.

*Figure-20. Variables tab in User Defined Feature Placement dialog box*

- If you want to modify any individual feature of UDF then click on the **Options** tab. The dialog box will be displayed as shown in Figure-21.

*Figure-21. Options tab in User Defined Feature Placement dialog box*

- If you want to scale up/down the dimensions of UDF which are not variable then click on the **Scale by value** radio button and specify the desired scale value.
- Select the desired radio button for dimensions that can not be varied while placing UDF. If you select the **Unlock** radio button then these dimensions can be edited manually after placing UDF. If you select the **Lock** radio button then the dimensions will be locked. If you select the **Hide** radio button then the dimensions will be hidden.
- If you want to redefine any feature like, you want to change the depth of extrude feature in UDF or you want to change a sketch then select the check box(es) for feature(s) which you want to redefine from the **Redefine these features** area of the dialog box and click on the **Edit Definition** button. The related contextual tabs will be displayed providing the options to redefine feature(s). Modify the features as required.
- If you want to exclude the last features of UDF then clear the **Auto Regeneration** check box and set the desired number of features upto which you want to create the UDF; refer to Figure-22.

*Figure-22. Modifying number of features*

- If you have features that can be flipped in orientation then click on the **Adjustments** tab and click on the **Flip** button.
- Click on the **OK** button from the dialog box to create the feature and exit the dialog box.

## SCALING MODEL

The **Scale Model** tool in the **Operations** panel is used to scale-up or scale-down the whole model. The procedure to use this tool is given next.

- Click on the **Scale Model** tool from the expanded **Operations** panel in the **Model** tab of the **Ribbon**. The **Scale Model** dialog box will be displayed; refer to Figure-23.

*Figure-23. Scale Model dialog box*

- Specify the desired value of scale in the edit box. You can specify the scale value in the range of 0.00001 to 10000. If you want to scale down the model then specify the value less than 1.
- If you want to change the absolute accuracy of model according to the scale value then select the **Scale absolute accuracy** check box. Here, accuracy means the lowest size that can be displayed properly in Creo Parametric. Accuracy plays important role in meshing the model and performing analysis.
- After specifying the value, click on the **OK** button. The model will be scaled accordingly.

## SETTING FEATURES READ-ONLY

Sometimes features get changed unintentionally while working on the model. To safeguard from such incidents, you can make the features read-only after creating them. The procedure to do so is given next.

- Select the feature up to which you want to make all the features read only. Click on the **Set as last read only** option from the **Read Only** cascading menu in the expanded **Operations** panel of the **Model** tab in the **Ribbon**. The features up to the selected feature will become read only; refer to Figure-24.

*Figure-24. Making features read only*

- If you want to make all the features of model read-only, then click on the **Set all as read only** tool from the **Read Only** cascading menu in the expanded **Operations** panel of the **Ribbon**.
- To clear read-only from feature, right-click on **Read Only Features** in the **Model Tree** and select the **Clear All Read Only** option; refer to Figure-25.

*Figure-25. Clear All Read Only option*

# FAMILY TABLE

The **Family Table** tool is used to create multiple instances of same part with some dimensional or feature variation. The procedure to create a family table is given next.

- Create a model of part for which you want to create family table; refer to Figure-26.
- Double-click on any flat face of the part to activate dynamic edit box. The dynamic editing mode will become active; refer to Figure-27. Or, you can right-click on the feature from **Model Tree** and select the **Edit dimension** button; refer to Figure-28. The dynamic editing mode will become active.

*Figure-26. Square nut for family table*

*Figure-27. Dynamic editing mode of part*

*Figure-28. Edit Dimension button*

- Select a dimension of the model that you want to include in the family table. The **Dimension** contextual tab will be displayed in the **Ribbon**; refer to Figure-29.

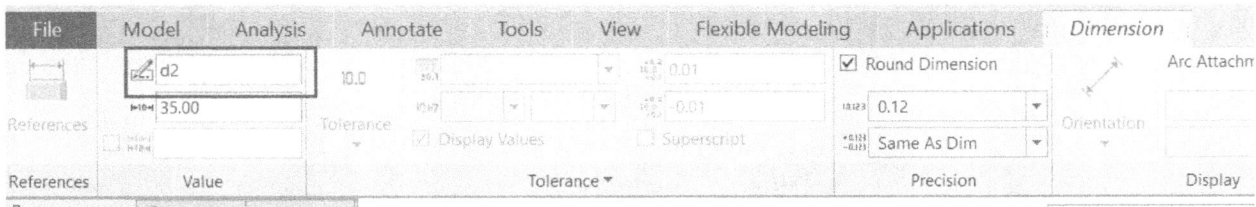

*Figure-29. Dimension contextual tab*

- Click in the **Modify symbol name of the dimensions** edit box highlighted in Figure-29 and enter the name by which you can identify the dimension in table like **Nut size**. Note that you can use alphabets, numerics and _ (Underscore) in specifying name.
- Similarly, set the names of other dimensions.
- Click on the **Family Table** tool from the expanded **Model Intent** panel in the **Model** tab of the **Ribbon**. The **Family Table** dialog box will be displayed; refer to Figure-30.

*Figure-30. Family Table dialog box*

- Click on the **Add/delete the table columns** button ⬚ from the toolbar in the dialog box. The **Family Items** dialog box will be displayed; refer to Figure-31. By default, the **Dimension** radio button is selected so you are asked to select dimensions to be included in the table.

*Figure-31. Family Items dialog box*

- Click on the feature of model whose dimensions are to be included in family table. The dimensions of feature will be displayed; refer to Figure-32.

*Figure-32. Dimensions of extrude feature displayed*

- Click one by one on the dimensions to be included in table; refer to Figure-33.

*Figure-33. Dimensions added in the table*

- If you want to add/remove a feature from the part using Yes/No option in the table then select the **Feature** radio button from the dialog box and select the features from the model. Similarly, you can include other items in the family table.
- Click on the **OK** button from the dialog box. The parameters of part will be displayed in the table; reefer to Figure-34.

*Figure-34. Parameters of part in family table*

- Click on the **Insert a new instance at the selected row** ⊞ button. A new instance of part will be displayed. Specify the desired name of instance.
- Specify the desired parameters for new instance and then press ENTER to create another instance.
- Repeat the process until you get required number of instances; refer to Figure-35. You can click on the **Edit with Excel** option from **File** menu in the dialog box to edit table in Microsoft Excel if you Microsoft Excel installed in your system.

| e | Instance Name | Common Name | F77 [HELICAL SWEEP_1] | d2 NUT_SIZE | d1 HOLE_DIA | d0 NUT_THICKNESS | d4 PITCH | d5 THREAD_DEPTH |
|---|---------------|-------------|------------------------|-------------|-------------|-------------------|----------|-----------------|
|   | NUT | nut.prt | Y | 35.000000 | 20.000000 | 10.000000 | 2.000000 | 2.000000 |
|   | NUT_D15 | nut.prt_INST | Y | 25 | 15 | 10 | 2 | 2 |
|   | NUT_D10 | nut.prt_INST | Y | 20 | 10 | 5 | 1 | 1 |
|   | NUT_D25 | nut.prt_INST | Y | 40 | 25 | 10 | 2 | 2 |
|   | NUT_D30 | nut.prt_INST | Y | 45 | 30 | 15 | 2 | 2 |

*Figure-35. Parameters specified for nut*

- After entering the desired parameters, click on the **Verify instances of the family** ☑ button from the toolbar in the dialog box. The **Family Tree** dialog box will be displayed; refer to Figure-36.

*Figure-36. Family Tree dialog box*

- Click on the **Verify** button from the dialog box. If verification status for all the instances are Success then close the dialog box and click on the **OK** button from the **Family Table** dialog box. If verification status for any of the instances is Failure then close the dialog box, change the parameters for instance and then again click on the **Verify** button from the **Family Table** dialog box. Once all the instances are successful, click on the **OK** button to create the table.
- Save the file using **Save** button from the **Quick Access Toolbar**. Close the file.

## Opening a Family Table Part

- Click on the **Open** button from the **Quick Access Toolbar**. The **File Open** dialog box will be displayed.
- Select the part for which family table was created earlier and click on the **Open** button. The **Select Instance** dialog box will be displayed; refer to Figure-37.

*Figure-37. Select Instance dialog box*

- Select the desired instance and click on the **Open** button. The selected instance of part will open.

## IMPORTING PART/ASSEMBLY

You can import parts/assemblies of third party CAD software by using the Import tool. The procedure to use **Import** tool is given next.

- Click on the **Import** tool from the expanded **Get Data** panel in the **Model** tab of the **Ribbon**. The **Open** dialog box will be displayed.
- Double-click on the third party CAD part/assembly file. The **File** dialog box will be displayed as shown in Figure-38.

*Figure-38. File dialog box*

- Select the desired radio button from **Import** type area of the dialog box. Like, select the **Geometry** radio button to import part as geometry or **Facet** radio button to import part as surfaces. If you are not sure what type of object is there in imported part then select the **Automatic** radio button.
- To modify the profile of importing part, click on the **Details** button from the dialog box. The **Import Profile** dialog box will be displayed as shown in Figure-39.

*Figure-39. Import Profile dialog box*

- Select the **Use templates** check box if you want to use Creo Parametric template while importing the part.
- Click on the **Surface** tab in the dialog box to modify parameters of the surface. The dialog box will be displayed as shown in Figure-40.

*Figure-40. Surface tab in the Import Profile dialog box*

- Select the desired parameters like select the **Keep b-spline surfaces** check box to keep b-spline surfaces in the imported model. In the same way, you can modify the other parameters of the profile.
- Click on the **OK** button from the **Import Profile** dialog box to apply changes.
- Click on the **OK** button from the dialog box. Preview of the imported part will be displayed; refer to Figure-41.

*Figure-41. Preview of imported part*

- Click on the **Import as solid** button ☐ from the **Ribbon** if you want to import the model as solid part. Click on the **Import as surface** button ☐ if you want to import the part as surface.
- Click on the **OK** button from the contextual tab.

# MERGE/INHERITANCE

The **Merge/Inheritance** tool is used to perform boolean operation of part with another part using the assembly constraints. The procedure is given next.

- Click on the **Merge/Inheritance** tool from the **Get Data** panel in the **Model** tab of the **Ribbon**. The **Merge/Inheritance** contextual tab will be displayed; refer to Figure-42.

Figure–42. Merge/Inheritance contextual tab

- Click on the **Open** button ⌷ from the contextual tab. The **Open** dialog box will be displayed.
- Double-click on the part file that you want to merge with the model in drawing area. Preview of the part will be displayed with **Component Placement** dialog box; refer to Figure-43.

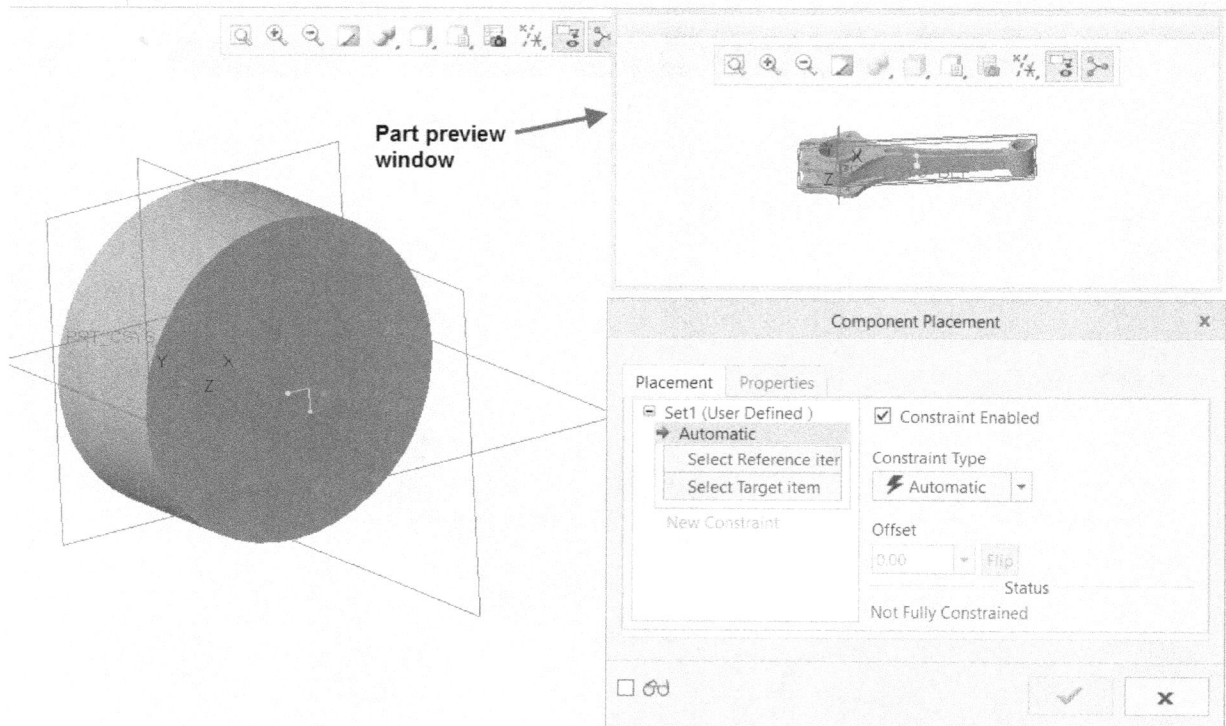

Figure–43. Part preview with Component Placement dialog box

- Select the faces of model in drawing area and part to be merged as required using the constraints. Note that the use of various constraints is discussed in Chapter 9 of this book. So, you can check the chapter and come back for practicing this tool.
- Apply desired constraints (We have used Default constraint which places the inserted part at default coordinate system) and click on the **OK** button from the dialog box. The buttons in contextual tab will become active and if you click on the **Verify** button ⌀ then you can check the preview of merge feature; refer to Figure-44. Click on the **Verify** button again to return.

*Figure-44. Preview of merge feature*

- Click on the **Toggle inheritance** button if you want to inherit all the parameters and features of the model to be merged. On doing so, you can later modify the features of inherited model using the **Model Tree**; refer to Figure-45.

*Figure-45. Inherited model features*

- To perform boolean operation, click on the desired button from the Ribbon. Click on the **Add Material** button [icon] to combine merged model with present model. Click on the **Remove Material** button [icon] to subtracted merged model from present model. Click on the **Intersect Material** button [icon] to create intersection part of the merging models.
- Click on the **OK** button from the contextual tab to create the feature.

## COLLAPSE

The **Collapse** tool is used to merge various features of the model to form one feature. In this way, you can hide the details of modeling features. The procedure to use this tool is given next.

• Select the features to you want to combine from the **Model Tree** and then click on the **Collapse** tool from the expanded **Editing** panel in the **Ribbon**. The **Collapse** dialog box will be displayed; refer to Figure-46.

*Figure-46. Collapse dialog box*

• Select the desired options from the dialog box and click on the **OK** button. An Incremental Geometry will be created by combining the selected features.

## RENDERING

Rendering is used to give realistic view to the part/assembly for presentation to the customer. For example, if your customer want to see how the assembly of watch created by you will look like in real life before he/she gives you order for production then it is better to present him the rendered image of product rather than starting the production directly. Creo Parametric has a very sophisticated system of photo realistic rendering. The tools for rendering are available in the **Render Studio** Application. The procedure to start the application is given next.

• After opening or creating the part/assembly model, click on the **Render Studio** tool from the **Rendering** panel in the **Applications** tab of the **Ribbon**. The **Render Studio** tab will be added in the **Ribbon**; refer to Figure-47. Also, the preview of render will be displayed; refer to Figure-47.

*Figure-47. Render Studio tab*

## Applying Appearances

- Click on the down arrow below **Appearance** tool and select the desired appearance from the list.
- If you want to apply material textures then click on the down arrow in the right side of **Library** node in the **Appearances box**. List of material categories will be displayed; refer to Figure-48.

*Figure-48. Material categories*

- Expand the category and select desired material type from the Material categories box. The related materials will be displayed in the **Library** node of **Appearances box**.
- Select the desired material. You will be asked to select the object to apply material texture.

- Select the **Part** filter from the **Selection Filter** drop-down at the bottom-right corner of the application window and select the part/face of the model.
- Press the Middle Mouse Button (MMB) to apply texture.

## User Defined Textures

If you want to use a texture download from internet or saved in your computer then follow the steps given next.

- Click on the down arrow below **Appearances** button in the **Project** panel of the **Render Studio** tab of the **Ribbon** and then select the **More Appearances** button from the **Appearances box**. The **Appearance Editor** dialog box will be displayed; refer to Figure-49.

- Specify the desired name for texture in the Name edit box.
- Set the desired class and sub-class for material using the drop-downs in the Properties tab of the dialog box and specify the other related parameters.
- Click on the **Texture** tab to edit the options related to texture of material. Select the **Image** option from the drop-down in the tab. The **Browse** button next to drop-down will become active.
- Click on the **Browse** button and select the texture of material download from internet or saved in your computer in image format.
- Similarly, you can add bump and decal to the material texture using the related tab options in the dialog box.
- Click on the **Close** button. The texture will be activated and your will be asked to select the object to apply texture. Select the object and press MMB to apply texture.
- Note that you can later find this texture in the class and sub-class that you have selected while creating the material texture.

*Figure-49. Appearance Editor dialog box*

## Applying Scenes

- Click on the down arrow below **Scenes** button in the **Project** panel of the **Render Studio** tab of the **Ribbon** and select the desired scene for rendering.

Similarly, select the desired orientation of part/assembly from the **Saved Orientations** drop-down in the **Ribbon**.

## Real-Time Rendering

Real-Time Rendering tool allows to check the rendering of model while applying textures to the model. To activate real-time rendering, click on the **Real-Time Rendering** toggle button from the **Real-Time** panel in the **Render Studio** tab of the **Ribbon** to set it as On. The rendering will be displayed. To edit the parameters of rendering follow the steps given next.

*   Click on the **Real-Time Settings** tool from the expanded **Real-Time** panel of the **Render Studio** tab of the **Ribbon**. The **Real-Time Rendering Settings** dialog box will be displayed; refer to Figure-50.

*Figure-50. Real-Time Rendering Settings dialog box*

*   Set the desired parameters in the dialog box or select desired preset from the **Lighting Presets** drop-down to system defined settings.
*   Click on the **OK** button from the dialog box to apply changes.

## Taking Screenshot

*   Click on the **Screenshot** tool from the **Render Output** panel in the **Render Studio** tab of the **Ribbon** to save the current rendering state of model as image file. The **Save screenshot** dialog box will be displayed; refer to Figure-51.

*Figure-51. Save screenshot dialog box*

- Specify a name for the file and click on the **Save** button to save the screen shot.

## Final Rendering

A good quality of render takes lots of computing power and time to render. So, it is advisable to first make changes during real-time rendering then move on to final render to save reattempts. The procedure for final rendering is given next.

- Click on the **Render** tool from the **Render Output** panel of the **Render Studio** tab of the **Ribbon**. The **Render** dialog box will be displayed; refer to Figure-52.

*Figure-52. Render dialog box*

- Specify the name and location of render file in the File Name edit box of the dialog box.
- Set the desired format for output file from the **Format** drop-down. For TIF and PNG format, you can also include transparency by select the **Include Alpha (Transparency)** check box next to drop-down.
- Set the desired resolution for output image file in the **Resolution** edit boxes.
- Set the termination condition of render from the Options area of the dialog box. You can specify maximum time or maximum samples for render termination.

- If you want to use the same settings as used for real-time rendering then select the **Use Real-Time Settings** check box otherwise clear the check box and expand the **Render Settings** node by click on the **+** button.
- The same options that were available in the **Real-Time Rendering Settings** dialog box will be displayed. Set the desired options and click on the **Render** button. The file will be saved at specified location after rendering.

# SELF ASSESSMENT

1. The datum graph is used to create a sketch graph which can be used in various relations and formulae in modeling. (T/F)

2. A User-Defined Feature is a combination of features that can be applied to part at one place. (T/F)

3. The .......... tool in the **Operations** panel is used to scale-up or scale-down the model.

4. The .......... tool is used to create multiple instances of same part with some dimensional or feature variation.

5. The .......... tool is used to perform boolean operation of part with another part using the assembly constraints.

6. The ............ tool is used to merge various features of the model to form one feature.

Answers to Self Assessment Questions :
1. T 2. T 3. **Scale Model** 4. **Family Table** 5. **Merge/Inheritance** 6. **Collapse**

# Chapter 9

# Assembly and Practical

### Topics Covered

The major topics covered in this chapter are:

- *Introduction*
- *Inserting Components in Assembly*
- *Assembly constraints and relationships*
- *Exploded View*
- *Copying components in assembly with constraint*
- *Practice*

# INTRODUCTION

Most of the things you find around you in real-world are assembly of various components; the computer you are working on is an assembly, the automotive you may be driving or sitting in for traveling is an assembly, there are lots of examples. Till this point, we have learned to create solid parts. But most of the time, we need to assemble the parts to get some use of them. In this chapter, we will work on the Assembly environment of Creo Parametric in which we will be assembling two or more components using some assembly constraints.

# STARTING ASSEMBLY ENVIRONMENT

The procedure to start assembly environment is given next.

- Click on the **New** button from the **Quick Access Toolbar** or **Ribbon**. The **New** dialog box will be displayed.
- Select the **Assembly** radio button from the **Type** area in the dialog box and **Design** radio button from the **Sub-type** area of the dialog box.
- Clear the **Use default template** check box and click on the **OK** button from the dialog box. The **New File Options** dialog box will be displayed.
- Select the **mmns asm design** option from the list box to make assembly in Millimeter Newton Second unit system; refer to Figure-1 and click on the **OK** button from the dialog box. The assembly environment will be displayed; refer to Figure-2.

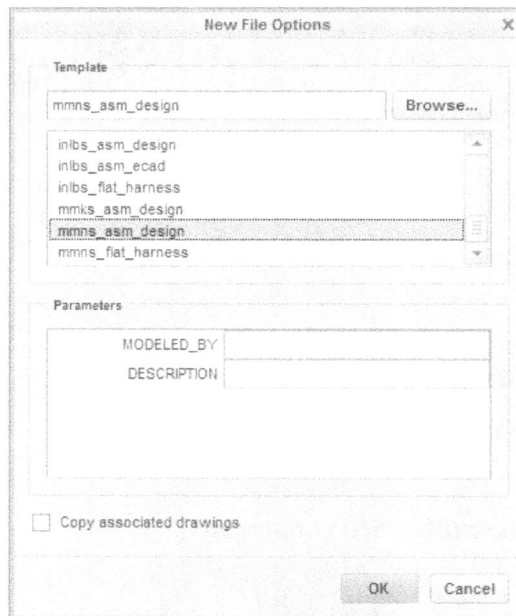

*Figure-1. New File Options dialog box*

*Figure-2. Assembly environment of Creo*

Note that you can change the units of assembly in the same way as discussed in Chapter 1. Changing the unit system will also affect the size of parts in the model.

## INSERTING BASE COMPONENT IN ASSEMBLY

In an assembly created in any CAD software, the first component is assumed to be the base component and it is generally fixed to its location. Afterwards, all the components are assembled with respect to the base component. The procedure of inserting components in assembly is given next.

• Click on the **Assemble** button from the **Assemble** drop-down in the **Component** panel of the **Ribbon**; refer to Figure-3.

*Figure-3. Assemble button*

• On doing so, the **Open** dialog box will be displayed; refer to Figure-4.

*Figure-4. Open dialog box*

- Select the base component from the dialog box and click on the **Open** button. The model will be displayed with a transformation triad and **Component Placement** contextual tab will be displayed; refer to Figure-5.

*Figure-5. Component placement*

- Sometimes, we may create the model in a different orientation than the required one in assembly. In such cases, hover the cursor on the desired arrow or ring of the triad and drag it in the desired direction; refer to Figure-6. Click on the **Show/ Hide 3D Dragger** button from the contextual tab to show/hide the triad.

*Figure-6. Transformation triad*

- After rotating the part, click on the **Relationship Type** drop-down from the contextual tab; refer to Figure-7. The list of constraints will be displayed.

*Figure-7. List of relationship constraints*

- Generally, either **Default** or **Fix** option is selected from the **Relationship Type** drop-down to place the base component. Select the desired constraint and click on the **OK** button from the contextual tab. During the chapter, you will get to know about each option in the contextual tab. Details of various constraints in **Relationship Type** drop-down is given next.

## CONSTRAINTS IN RELATIONSHIP TYPE DROP-DOWN

There are 11 options available in the **Relationship Type** drop-down to place components in the assembly. These options are discussed next.

## Distance

As the name suggests, this constraint is used to place the component at a specified distance from the selected reference. The procedure to use this option is given next.

- Click on the **Distance** option from the **Relationship Type** drop-down in the contextual tab or from the **Constraint Type** drop-down in the **Placement** tab; refer to Figure-8. You are asked to select one reference from component and one reference from assembly.

*Figure-8. Distance constraint*

- Select a face/plane/edge/axis/point from the component. You are asked to select a reference from assembly. Also a cosmetic line gets attached to the cursor.
- Select a face/plane/edge/axis/point of assembly. The distance constraint will be applied and the distance will be displayed as a dimension; refer to Figure-9.

*Figure-9. Selecting faces for distance constraint*

- Double-click on the dimension value or specify the desired distance value in the **Offset** edit box; refer to Figure-10.

*Figure-10. Distance value*

- Click on the **Flip** button next to the **Offset** edit box to flip the component in offset direction.
- To apply next constraint, click on the **New Constraint** option from the **Placement** tab; refer to Figure-11.

*Figure-11. New Constraint option*

## Angle Offset

As the name suggests, this constraint is used to set an angle between the selected planes/faces/edges of component and assembly. The procedure to use this constraint is given next.

- Click on the **Angle Offset** option from the **Relationship Type** drop-down in the contextual tab or from the **Constraint Type** drop-down in the **Placement** tab. You are asked to select one reference from component and one reference from assembly.
- Select the face/plane/edge/axis of the component and then of the assembly. The angle offset constraint will be applied; refer to Figure-12.
- Specify the desired angle value. To apply the next constraint, click on the **New Constraint** option from the **Placement** tab.

*Figure-12. Angle offset constraint applied*

## Parallel

As the name suggests, this constraint is used to set the component and assembly parallel for selected faces/planes/edges/axes. The procedure to use this constraint is given next.

- Click on the **Parallel** option from the **Relationship Type** drop-down in the contextual tab or from the **Constraint Type** drop-down in the **Placement** tab. You are asked to select one reference from component and one reference from assembly.
- Select the face/plane/edge/axis of the component and then of the assembly. The component will become parallel to the assembly with respect to the selected references; refer to Figure-13.

Before applying parallel constraint

After applying parallel constraint

*Figure-13. Applying parallel constraint*

## Coincident

Coincident means, to be at the same place or position. Using this constraint, we can make references(selected entities of assembly and component) lie at the same place or position. The procedure to use this constraint is given next.

- Click on the **Coincident** option from the **Relationship Type** drop-down in the contextual tab or from the **Constraint Type** drop-down in the **Placement** tab. You are asked to select one reference from component and one reference from assembly.
- Select the face/plane/edge/axis/point of the component and then of the assembly. The component will become coincident at the selected reference; refer to Figure-14.

*Figure-14. Applying coincident constraint*

## Normal

Normal constraint is used to make references(selected entities of assembly and component) perpendicular to each other. The procedure to use this constraint is given next.

- Click on the **Normal** option from the **Relationship Type** drop-down in the contextual tab or from the **Constraint Type** drop-down in the **Placement** tab. You are asked to select one reference from component and one reference from assembly.
- Select the face/plane/edge/axis of the component and then of the assembly. The component will become perpendicular to assembly at the selected reference; refer to Figure-15.

*Figure-15. Applying Normal constraint*

## Coplanar

Coplanar constraint is used to make references(selected entities of assembly and component) share the same plane. The procedure to use this constraint is given next.

- Click on the **Coplanar** option from the **Relationship Type** drop-down in the contextual tab or from the **Constraint Type** drop-down in the **Placement** tab. You are asked to select one reference from component and one reference from assembly.
- Select the axis of the component and then of the assembly. The component will become coplanar to assembly at the selected reference; refer to Figure-16.

*Figure-16. Applying Coplanar constraint*

## Centered

Centered constraint is used to make component centered to the assembly with respect to the selected round face. The procedure to use this constraint is given next.

- Click on the **Centered** option from the **Relationship Type** drop-down in the contextual tab or from the **Constraint Type** drop-down in the **Placement** tab. You are asked to select one reference from component and one reference from assembly.
- Select a round face of the component and then of the assembly. The component will become centered to assembly at the selected reference; refer to Figure-17.

*Figure-17. Applying centered constraint*

## Tangent

Tangent constraint is used to make component tangent to the assembly with respect to the selected round face. The procedure to use this constraint is given next.

- Click on the **Tangent** option from the **Relationship Type** drop-down in the contextual tab or from the **Constraint Type** drop-down in the **Placement** tab. You are asked to select one reference from component and one reference from assembly.
- Select a round face of the component and then of the assembly. The component will become centered to assembly at the selected reference; refer to Figure-18.

*Figure-18. Applying tangent constraint*

## Fix

As the name suggests, Fix constrain is used fix the component at its present location. This constraint is generally used to place the base component of the assembly. The procedure to use this constraint is given next.

- Click on the **Fix** option from the **Relationship Type** drop-down in the contextual tab or from the **Constraint Type** drop-down in the **Placement** tab. The component will be fixed at its current location. Note that when the component is fully defined, its color changes from violet to dark orange. It shows that the component cannot move from its location due to any motion-simulation relation; refer to Figure-19.

*Figure-19. Component fixed at its location*

## Default

As the name suggests, Default constrain is used place the component at its default position in assembly. It means- Front plane of component will be coincident with Front plane of assembly, Right plane of component will be coincident with the Right plane of assembly and similarly, the Top plane of component will be coincident with the Top plane of assembly. Primarily, this constraint is also used to place the base component of the assembly. The procedure to use this constraint is given next.

- Click on the **Default** option from the **Relationship Type** drop-down in the contextual tab or from the **Constraint Type** drop-down in the **Placement** tab. The component will be placed according to the default planes of assembly; refer to Figure-20.

Before applying constraint       After applying constraint

*Figure-20. Applying default constraint*

## CONSTRAINTS IN MECHANISM CONNECTIONS DROP-DOWN

From the name Mechanism, you can have a fair idea that this drop-down has something to do with mechanism. Using the options in the **Mechanism Connections** drop-down, we can constrain the component in such a way that the connection reacts as real-machine mechanism; refer to Figure-21. Like, we can create a sliding joint between the two components in assembly. The constraints in this drop-down are discussed next.

*Figure-21. Mechanism Connections drop-down*

## Rigid

As the name suggests, **Rigid** option is used to apply rigid connection between two components of an assembly. Select this option and you are allowed to use all the constraints that are available in the **Relationship Type** drop-down.

## Pin

**Pin** constraint is used to assemble the component in such a way that they act like hinged at the selected reference. The procedure to use this constraint is given next.

- After inserting component, click on the **Pin** option from the **Mechanism Connection** drop-down; refer to Figure-22.
- Click on the **Placement** tab and check the requirements under **Connections** tab. The first requirement of constraint is axis alignment.
- Select an axis of component and then of the assembly. The axes will be aligned; refer to Figure-23. You are asked to select references for restricting the translation of component.

Figure-22. Pin constraint

Figure-23. Axes alignment

- Select a face/plane/vertex of component and then of the assembly. The coincident relation will be applied to restrict the translation; refer to Figure-24 and Figure-25.

*Figure-24. Restricting translation of component*

*Figure-25. Applying pin constraint*

- Using the Blue ring in 3D dragger (so called triad), you can rotate the hinged plate.
- Click on the **Rotation Axis** option from the **Placement** tab and select one face of each component and assembly; refer to Figure-26.

*Figure-26. Rotation limit for pin*

- Click in the **Current Position** edit box and specify the value as **0**. Select the **Minimum Limit** and **Maximum Limit** check boxes and specify the rotation limits.
- Click on the **OK** button from the contextual tab to apply the constraint.

## Slider

**Slider** constraint is used to assemble the component in such a way that it acts like hinged at the selected reference. The procedure to use this constraint is given next.

- After inserting component, click on the **Slider** option from the **Mechanism Connection** drop-down; refer to Figure-27.

Figure-27. Slider constraint

- Click on the **Placement** tab and check the requirements under **Connections** tab. The first requirement of constraint is axis alignment.
- Select an axis of component and then of the assembly. The axes will be aligned; refer to Figure-28. You are asked to select references for restricting the rotation of component.

Figure-28. Axes aligning

- Select face/plane of component and then of assembly; refer to Figure-29.

Figure-29. Restricting rotation of part

- Click on the **Translation Axis** option from the **Placement** tab. You are asked to select reference for translation limits; refer to Figure-30.

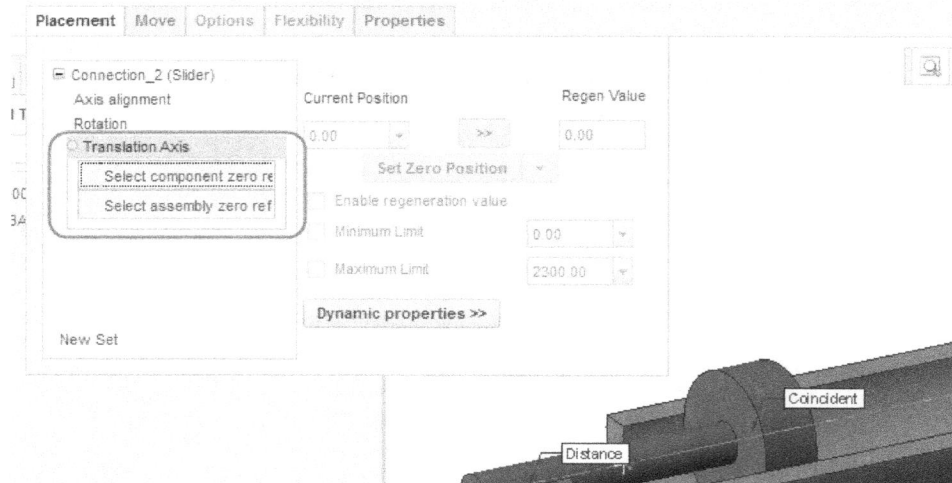

*Figure-30. Translation Axis option*

- Select face of component and then of the assembly; refer to Figure-31.

*Figure-31. Faces selected for translation*

- Click in the **Current Position** edit box and specify the value as **0** if you want to match the two faces for defining starting.
- Select the **Minimum Limit** and **Maximum Limit** check boxes and specify the values respectively.
- After specifying the desired values, click on the **OK** button from the contextual tab to create the connection.

## Confirm Motion of Components

Once you have applied a connection then you can check it by using the **Drag Components** button. The procedure for checking motion of slider in previous example is given next.

- Click on the **Drag Components** button from the **Ribbon**. You are asked to select a component.
- Click on the slider(Component) and move the cursor. The component will be able to slide on the base only.

## Cylinder

**Cylinder** constraint is used to assemble the component in such a way that the component is free to move along the selected axis and free to rotate about the same selected axis. But, you can specify the limits for rotation as well as translation. The procedure to use this constraint is given next.

- After inserting component, click on the **Cylinder** option from the **Mechanism Connection** drop-down.
- Click on the **Placement** tab from the contextual tab. You are asked to select axis for alignment.
- Select an axis of component and then of the assembly.
- Specify the desired limitations for translation and rotation. Refer to Figure-32.

*Figure-32. Cylinder constraint*

## Planar

**Planar** constraint is used to assemble the component in such a way that the component can move along the selected plane in defined boundary. You need to specify the limits for rotation and translation. The procedure to use this constraint is given next.

- After inserting component, click on the **Planar** option from the **Mechanism Connection** drop-down.
- Click on the **Placement** tab from the contextual tab. You are asked to select a face/plane for applying coincident relation.
- Select a face of component and then of the assembly.
- Specify the desired limitations for translation and rotation. Figure-33 shows an application of Planar constraint

*Figure-33. Planar constraint*

# Ball

**Ball** constraint is used to assemble the component in such a way that the component is free to move 360 degree in 3D space pivoted to a point. This kind of connection is required when there is a ball joint in the assembly. The procedure to use this constraint is given next.

- After inserting component, click on the **Ball** option from the **Mechanism Connection** drop-down.
- Click on the **Placement** tab from the contextual tab. You are asked to select a point/vertex for applying coincident relation.
- Select a point of component and then of the assembly; refer to Figure-34.
- Click on the **Cone Axis** option to specify limitations to rotation of ball joint. You are asked to select axis of the component.
- Select the center axis of the ball. You are asked to select a reference point on the assembly.
- Select a point of the assembly. Preview of cone will be displayed; refer to Figure-35.

*Figure-34. Point selection for ball joint*

*Figure-35. Preview of rotation limiting cone*

- Specify the opening angle of the cone as shown in above figure. Note that you may need to select the **Flip** button from the **Placement** tab to get such a cone.
- Click on the **OK** button from the contextual tab to create the joint.
- Use the **Drag Components** button to check motion of the ball joint.

## Weld

**Weld** constraint is used to permanently join component to the assembly. This constraint is same like fixing a component at a place. But in this case, the component will move along with the assembly. The procedure to use this constraint is given next.

- After inserting component, click on the **Weld** option from the **Mechanism Connection** drop-down.
- Click on the **Placement** tab from the contextual tab. You are asked to select a coordinate systems.
- Select the coordinate system of component and then of the assembly. The color component will change to orange showing that the component is fully placed.

## Bearing

**Bearing** constraint is used to apply the same motion to the component which is applied on a ball in the ball bearing. After applying Bearing constraint, the component can rotate in any direction but can move only along the selected direction. This constraint is a combination of slider and ball constraints discussed earlier. The procedure to use this constraint is given next.

- After inserting component, click on the **Bearing** option from the **Mechanism Connection** drop-down.
- Click on the **Placement** tab from the contextual tab. You are asked to select a point/vertex of the component.
- Select the point component. You are asked to select an edge/axis of the assembly.
- Select an edge or axis of the assembly.
- Specify the translation and rotation limits as discussed earlier in this chapter.
- Click on the **OK** button from the contextual tab to apply the constraint.

## General

**General** constraint is a kind of master tool for constraining the components. Using the **General** constraint, you can apply any of the joint/connection we have discussed earlier. The procedure to use this constraint is given next.

- After inserting component, click on the **General** option from the **Mechanism Connection** drop-down.
- Click on the **Placement** tab from the contextual tab. We have three constraints to apply, primarily; refer to Figure-36.

*Figure-36. General constraint*

- Select one reference from component and one reference from assembly, and then apply the desired relation from the available three constraints. You are asked to either specify more constraints or you can specify translational/rotational limits; refer to Figure-37.

*Figure-37. Applying General constraint*

- Set the desired option and click on the **OK** button from the contextual tab to apply the constraint.

## 6DOF

**6DOF** (6 Degree of Freedom) constraint is used to specify the limits of movement of component in each direction. Using this constraint, you can not restrict the component rotation. Procedure to apply this constraint is similar to **Weld** constraint but in this case, component is free to move.

## Gimbal

**Gimbal** constraint is used to make the component centered to the assembly with respect to the selected coordinate system. Procedure to apply this constraint is similar to **Weld** constraint but in this case, component is free to rotate.

## Slot

**Slot** constraint is used to make the component follow path created on the assembly component. The procedure to use this constraint is given next.

- After inserting component, click on the **Slot** option from the **Mechanism Connection** drop-down.
- Click on the **Placement** tab from the contextual tab. You are asked to select a vertex or point of the component.
- Select the vertex of the component; refer to Figure-38. You are asked to select an edge/axis/curve along which the follower will move.

*Figure-38. Vertex selected*

- Select the sketched curve/edge/axis along which you want to make the follower move; refer to Figure-39.

*Figure-39. Curve selected for path*

- Click on the **OK** button from the contextual tab to apply the constraint. Using the **Drag Components** tool you can check the motion of the component.
- You can apply more constraints to the follower so that it remain perpendicular to the face of the assembly base. Like, you can use the **General** constraint after selecting **New Set** option in the **Placement** tab and use the Parallel relationship.

# EXPLODED VIEW

We have learned about assembling the component till now. But, on paper, we need to display the components of assembly in such a way that the other person can easily find out the location of a component in assembly. For this purpose, we create an exploded view of assembly. In exploded view, the components are disassembled and placed at some distance from each other according to their sequence of assembling. The procedure to create exploded view of an assembly is given next.

- Open any assembly earlier created or create an assembly.
- Click on the **Exploded View** button from the **Model Display** panel of the **Ribbon**. The default exploded view of the assembly will be displayed; refer to Figure-40.

*Figure-40. Default exploded view*

- The exploded view displayed in the above figure is created by placing each component at default specified distance from other. Also, the components are displaced as per their assembly sequence. We suggest you to maintain the same assembly sequence although you can change the distance between them for a clear view.
- To change the distance between components, click on the **Edit position** button from the **Model Display** panel in the **Ribbon**. The **Explode Tool** contextual tab will be displayed.

- Click on the component that you want to move. A triad will be displayed with the component allowing you to displace the component; refer to Figure-41.

*Figure-41. Component to be moved*

- Click on an arrow and drag the component to the desired location. Similarly, you can displace the other components of the assembly.
- You can also apply cosmetic lines to show the connection of components to each other in the assembly. To do so, click on the **Cosmetic Lines** button from the contextual tab. The **Cosmetic Offset Line** dialog box will be displayed; refer to Figure-42 and you are asked to select two references.

*Figure-42. Cosmetic Offset Line dialog box*

- Click any two references of the components between which you want to create the cosmetic line; refer to Figure-43.

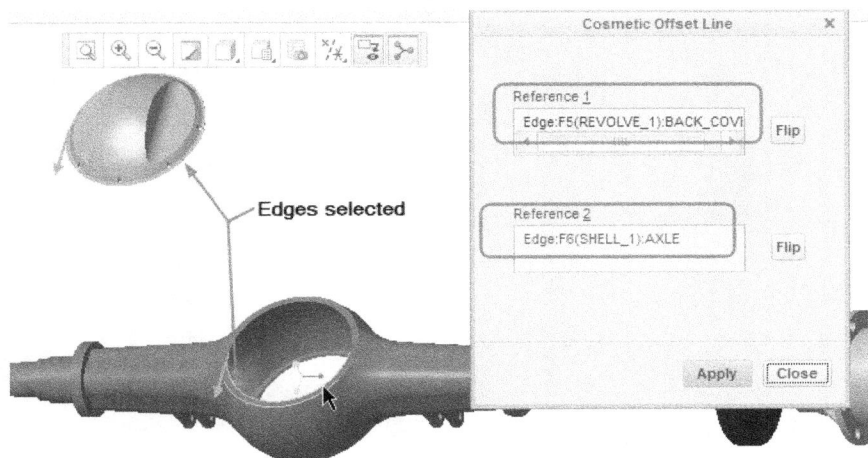

*Figure-43. References selected cosmetic line*

- Click on the **Apply** button to create the cosmetic line; refer to Figure-44.

*Figure-44. Cosmetic line created*

- Continue the process till you have desired cosmetic lines to show the connections between components.
- Click on the **Close** button to exit the **Cosmetic Offset Line** dialog box.
- Click on the **OK** button from the contextual tab to apply the exploded view.
- To return back to assembled view of assembly, click again on the **Exploded View** button.

## VIEW MANAGER AND ITS ROLE

Recently, we have created an exploded view of the assembly. You can create more than one exploded view of the assembly and you can also create sections of the assembly to see inside of the assembly. These all options are available in **View Manager**. The procedure to create various views using the **View Manager** is given next.

- Click on the **View Manager** button from the **Manage Views** drop-down in the **Ribbon**; refer to Figure-45. The **View Manager** dialog box will be displayed; refer to Figure-46.

*Figure-45. View Manager button*

*Figure-46. View Manager dialog box*

- Click on the **Explode** tab in the dialog box if not selected by default.
- Click on the **New** button from the dialog box to create a new exploded view.
- Specify the name of the view as desired and press the Middle click button. Using the **Edit Position** button from the **Ribbon**, change the position of components.
- Right-click on the newly create exploded view; refer to Figure-47.

*Figure-47. View Manager with newly created exploded view*

- Click on the **Save** button from the shortcut menu and then click on the **OK** button from the **Save Display Elements** dialog box displayed.
- To switch between various exploded views, right-click on the desired view in the **View Manager** dialog box and select the **Activate** option.

# APPEARANCE

For most of the parts, appearance is automatically applied when a material is assigned to them. But if you want to change the appearance of a part in assembly then the options are available in the **Appearances** drop-down; refer to Figure-48. The procedure to apply appearances is given next.

- Click on the down arrow for **Appearances** option in the **Model Display** panel of the **Model** tab in the **Ribbon**. A drop-down will be displayed as shown in Figure-48.
- Click on the desired color/appearance from the drop-down. You will be asked to select the item to which appearance is to be applied.
- Click on the part to apply the appearance to whole part. Hold the **CTRL** key to select multiple parts. If you want to select some portions of part in place of whole part then select **Surface** from the **Selection Filter** drop-down at the bottom-right corner of the application window and then select multiple portions of part while holding the **CTRL** key; refer to Figure-49.

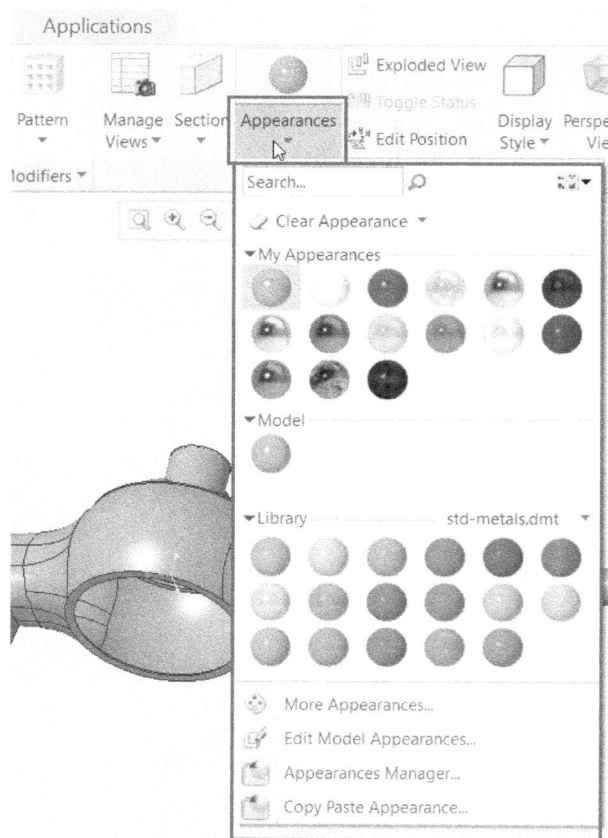

*Figure-48. Appearances drop-down*

*Figure-49. Selection Filter drop-down*

- After selection, press the Middle Mouse Button. The selected portion will be colored accordingly; refer to Figure-50.

*Figure-50. Applying appearances*

To remove appearances, click on the down arrow next to **Clear Appearance** option and select the desired option from the drop-down displayed; refer to Figure-51. To remove desired appearance, select the faces having the appearance and click on the **Clear Appearance** option from the drop-down. To remove all assembly appearances, click on the **Clear Assembly Appearances** option from the drop-down. Note that on using this option, the appearances applied in Part environment will not be removed. Click on the **Clear All Appearances** option to remove all the appearances. More advanced options of appearance are out of scope of this book.

*Figure-51. Clear Appearance drop-down*

# PRACTICAL 1

Create the real axle assembly as shown in Figure-52. The exploded view of assembly is given in Figure-53.

*Figure-52. Rear axle assembly*

| Sr. No. | Name | Quantity |
|---|---|---|
| 1 | AXLE | 1 |
| 2 | BACK_COVER | 1 |
| 3 | DISC | 2 |
| 4 | FRONT_COVER | 1 |
| 5 | SHAFT | 2 |
| 6 | SPINDLE | 2 |
| 7 | WHEEL_RIM | 2 |

*Figure-53. Exploded view of the assembly*

## Starting assembly file

- Start Creo Parametric and click on the **New** button from the **Quick Access toolbar**. The **New** dialog box will be displayed.
- Click on the **Assembly** radio button from the **Type** area and **Design** option from the **Sub-type** area.
- Specify desired name of the file.
- Clear the **Use default template** check box and click on the **OK** button from the dialog box. The **New File Options** dialog box will be displayed; refer to Figure-54.

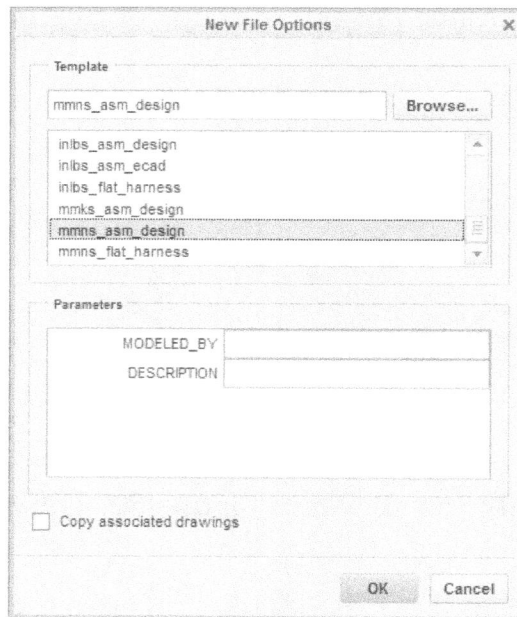

*Figure-54. New File Options dialog box*

- Select the **mmns asm design** option and click on the **OK** button.

## Inserting the Base Component (Axle)

- Click on the **Assemble** button from the **Assemble** drop-down in the **Components** panel of the **Ribbon**. The **Open** dialog box will be displayed; refer to Figure-55.

Note that if preview of component is not displaying in your case, then click on the Preview button at right corner of the dialog box. The preview window will be displayed.

Move the cursor in the preview window and scroll up/down to check the preview. Press the middle button of mouse and drag to rotate the component.

*Figure-55. Open dialog box*

- Select the part file of axle from dialog box and click on the **Open** button. The axle will be displayed in the assembly window; refer to Figure-56.

*Figure-56. Preview of axle*

- Select the **Default** option from the **Relationship type** drop-down in the **Component Placement** contextual tab; refer to Figure-57.

*Figure-57. Placing axle*

- Click on the **OK** button from the contextual tab to place the axle.

## Inserting the Front Cover

- Click on the **Assemble** button from the **Assemble** drop-down in the **Components** panel of the **Ribbon**. The **Open** dialog box will be displayed.
- Select the part file of **Front Cover** and click on the **Open** button from the dialog box.
- Hide the display of planes using the **Datum Display Filters** drop-down in the **In-Graphics** toolbar; refer to Figure-58.

*Figure-58. Hiding planes and 3D Dragger*

- Select the **Coincident** relation from the **Relationship type** drop-down.
- Select the center axis of component and then of assembly; refer to Figure-59.

*Figure-59. Axes selected*

- Click on the **Flip axis constraint** button ⤬ from the contextual tab to flip the component if required.
- Click on the **Placement** tab and click on the **New Constraint** option.
- Select the Coincident option from the Constraint Type drop-down in the Placement tab.
- Select the flat face of Front cover and Axle; refer to Figure-60. The component will be fully constrained.

*Figure-60. Faces selected for coincident constraint*

- Click on the **OK** button from the contextual tab.

## Inserting the Front Cover

- Click on the **Assemble** button from the **Assemble** drop-down in the **Components** panel of the **Ribbon**. The **Open** dialog box will be displayed.
- Select the part file of **Back Cover** and click on the **Open** button from the dialog box.
- Click on the **Placement** tab and select the **Coincident** option from the **Constraint Type** drop-down.
- Select the flat face of component and then of assembly; refer to Figure-61.

*Figure-61. Select flat faces*

- Click on the **New Constraint** option from the **Placement** tab and select the **Coincident** option from the **Constraint Type** drop-down.
- One by one select the axis of Back plate and Axle; refer to Figure-62.

*Figure-62. Axes selected for coincident constraint*

- Click on the **OK** button from the contextual tab to apply the constraint.

## Inserting the Disc

- Click on the **Assemble** button from the **Assemble** drop-down in the **Components** panel of the **Ribbon**. The **Open** dialog box will be displayed.
- Select the part file of **Disc** and click on the **Open** button from the dialog box.
- Click on the **Placement** tab and select the **Coincident** option from the **Constraint Type** drop-down.
- Select the flat face of component and then of assembly; refer to Figure-63.

*Figure-63. Selecting flat faces of disc and axle*

• Click on the **New Constraint** option from the **Placement** tab and select the **Coincident** option from the **Constraint Type** drop-down.
• One by one select the axis of Disc and Axle; refer to Figure-64.

*Figure-64. Selecting axes*

• Click on the **OK** button from the contextual tab to apply the constraints.

## Inserting the Spindle

• Click on the **Assemble** button from the **Assemble** drop-down in the **Components** panel of the **Ribbon**. The **Open** dialog box will be displayed.
• Select the part file of **Spindle** and click on the **Open** button from the dialog box.
• Click on the **Placement** tab and select the **Coincident** option from the **Constraint Type** drop-down.
• Select the center axis of component and then of assembly; refer to Figure-65.

*Figure-65. Axes of spindle and axle*

- Click on the **New Constraint** option from the **Placement** tab and select the **Coincident** option from the **Constraint Type** drop-down.
- One by one select the faces of Disc and Axle; refer to Figure-66.

*Figure-66. Faces of spindle and axle selected*

- Click on the **OK** button from the contextual tab to apply the constraint.

## Inserting the Wheel Rim

- Click on the **Assemble** button from the **Assemble** drop-down in the **Components** panel of the **Ribbon**. The **Open** dialog box will be displayed.
- Select the part file of **Wheel rim** and click on the **Open** button from the dialog box.
- Click on the **Placement** tab and select the **Coincident** option from the **Constraint Type** drop-down.
- Select the center axis of component and then of assembly; refer to Figure-67.

*Figure-67. Axes alignment of wheel rim*

- Click on the **New Constraint** option from the **Placement** tab and select the **Coincident** option from the **Constraint Type** drop-down.
- Select the hole axis of Spindle and then of the Wheel Rim; refer to Figure-68.

*Figure-68. Axes of holes aligned*

- Again, click on the **New Constraint** option from the **Placement** tab and select the **Coincident** option from the **Constraint Type** drop-down.
- Select the flat faces of Spindle and Wheel Rim; refer to Figure-69.

*Figure-69. Flat faces selected*

## Inserting the Wheel Rim

- Click on the **Assemble** button from the **Assemble** drop-down in the **Components** panel of the **Ribbon**. The **Open** dialog box will be displayed.
- Select the part file of **Wheel rim** and click on the **Open** button from the dialog box.
- Click on the **Placement** tab and select the **Coincident** option from the **Constraint Type** drop-down.
- Select the center axis of component and then of assembly; refer to Figure-70.

*Figure-70. Aligning axis of shaft*

- Click on the **New Constraint** option from the **Placement** tab and select the **Coincident** option from the **Constraint Type** drop-down.
- Select the flat faces of Shaft and Wheel Rim; refer to Figure-71.

*Figure-71. Faces of wheel rim and shaft selected*

We have done the assembly on one side. Now, we will learn how to create copy of components in assembly with assembly constraints.

## Copying Components in Assembly

- Select Disc, Spindle, Wheel Rim and Shaft from the Model Tree while holding the CTRL key; refer to Figure-72.
- Hold the RMB. A shortcut menu will be displayed.
- Select the **Copy** option from the shortcut menu; refer to Figure-73.

*Figure-72. Components selected from Model Tree*

*Figure-73. Copy option*

- Right-click anywhere in the screen and select the **Paste option with dependency configuration**; refer to Figure-74. The **Paste Special** dialog box will be displayed; refer to Figure-75.

Figure-74. *Paste option with dependency configurations*

Figure-75. *Paste Special dialog box*

- Select the **Advanced reference configuration** check box from the dialog box and then click on the **OK** button from the dialog box. The **Advanced Reference Configuration** dialog box will be displayed; refer to Figure-76.

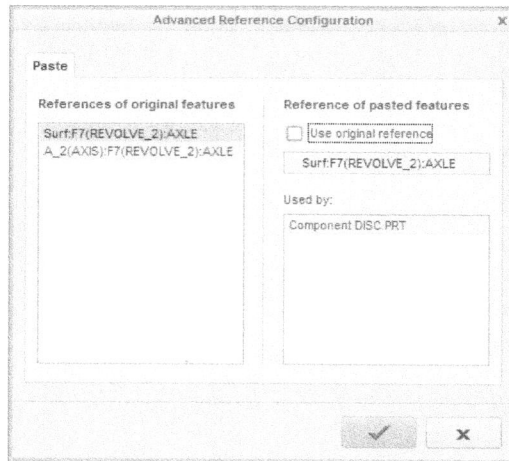

Figure-76. *Advanced Reference Configuration dialog box*

- Through this dialog box, you are asked to specify the replacement references for the reference entities displayed in the left list box. By default, the Surf:F7(REVOLVE 2):AXLE reference surface is selected in the left box.
- Select the same face on the other side of the axle; refer to Figure-77. You are asked to select the replacement of axis.

*Figure-77. Face replaced*

• Select the same axis on the other side of the axle; refer to Figure-78.

*Figure-78. Axis replaced*

• Click on the **OK** button from the dialog box. The copy will be created with the same constraints on other side; refer to Figure-79.

*Figure-79. Assembly after creating copy*

# PRACTICE 1

Assemble the model of tail stock as shown in Figure-80. The exploded view of tail stock is given in Figure-81. The Bill of Material in given in Figure-82. Note that you will learn about Bill of Material later in the Book.

*Figure-80. Practice 1*

*Figure-81. Exploded view of tail stock*

| Sr. No. | Component Name | Qty. |
|---|---|---|
| 1 | BARREL----3 | 1 |
| 2 | BODY-----1 | 1 |
| 3 | CENTRE----19 | 1 |
| 4 | CLAMPING-PLATE----15 | 1 |
| 5 | FEATHER--2 | 1 |
| 6 | FEATHER-KEY--7 | 1 |
| 7 | FLANGE---5 | 1 |
| 8 | HAND-WHEEL-------8 | 1 |
| 9 | HANDLE---13 | 1 |
| 10 | HEXAGONAL---NUT--M16--------14 | 1 |
| 11 | HEXAGONAL---NUT-M12------10 | 1 |
| 12 | HEXAGONAL-NUT-M22--------18 | 1 |
| 13 | SCREW---6 | 4 |
| 14 | SCREW--SPINDILE---4 | 1 |
| 15 | SQ-HEADED-BOLT--16 | 1 |
| 16 | STUD---11 | 1 |
| 17 | WASHER----M12------9 | 1 |
| 18 | WASHER----M22-STD----------17 | 1 |
| 19 | WASHER-M16---------12 | 2 |

*Figure-82. Bill of Material*

# SELF ASSESSMENT

1. The .......... constraint is used to set the component and assembly parallel for selected faces/planes/edges/axes.

2. ............. constraint is used to make references(selected entities of assembly and component) share the same plane.

3. ........... constrain is used fix the component at its present location.

4. ............... constraint is used to assemble the component in such a way that they act like hinged at the selected reference.

5. .............. constraint is used to assemble the component in such a way that the component is free to move 360 degree in 3D space pivoted to a point.

6. Which of the following constraints is used to permanently join component to the assembly?

(a) Weld          (b) Fix
(c) Pin           (d) Coplanar

7. Which of the following constraints is used to make the component follow path created on the assembly component?

(a) Pin           (b) Slot
(c) Coplanar      (d) Gimbal

**Answers to Self Assessment Questions :**

1. **Parallel**  2.  **Coplanar**  3. **Fix**  4. **Pin**  5. **Ball**  6. (a)  7. (b)

# Chapter 10

# Advanced Assembly
# and
# Introduction to Mechanism

## Topics Covered

The major topics covered in this chapter are:

- *Introduction*
- *Bottom-Up Assembly Approach*
- *Modifying Part in Assembly*
- *Mechanism Design*

Driven disk

Driver disk

# INTRODUCTION

In previous chapter, you have learned about inserting components in assembly and performing simple assembly operations. In this chapter, you will learn about creating parts in the assembly (also called Bottom-Up Assembly approach), performing mechanism simulations, and many other advanced operations. We will start the chapter with Bottom-Up Assembly approach.

# BOTTOM-UP ASSEMBLY APPROACH

In the previous chapter, we inserted the components in assembly file which were already created. Now, we will create components as per the requirement of assembly. For this we have opened an already created assembly in Creo Parametric as shown in Figure-1. In this assembly, we will create a lever that can join the boss features of two sliders as shown in Figure-2. Before that we will change the shuttle part shown in the figure because it is overlapping with the base part. The procedure is discussed next.

*Figure-1. Assembly for Bottom Up Approach*

*Figure-2. Lever to be created*

## Modifying Part in Assembly

- Open the assembly with the name Bottom-Up Approach.asm from the resources available for this book in Chapter 10. The assembly will be displayed as shown in Figure-1.
- Right-click on any of the Shuttle.prt from **Model Tree** and click on the **Activate** button; refer to Figure-3. The part will be activated in drawing area; refer to Figure-4.

*Figure-3. Activate button*

*Figure-4. Shuttle part activated in assembly*

- Click on the right face of the shuttle part. Mini Toolbar will be displayed; refer to Figure-5.

*Figure-5. Mini toolbar on selecting face of part*

- Click on the **Edit Definition** button from the Mini toolbar. The **Extrude** contextual tab will be displayed and part will change to editing mode.
- Click on the **Placement** tab and then click on the **Edit** button. The sketch editing mode will become active; refer to Figure-6.

*Figure-6. Editing sketch of shuttle*

- Delete the dimensions of sketch and make the outer edge of shuttle coincident with edge of groove in base; refer to Figure-7. Similarly, make the top line coincident with upper edge; refer to Figure-8.

*Figure-7. Making line coincident*

*Figure-8. Sketch after making coincident*

- Click on the **OK** button from the **Sketch** contextual tab and then from the **Extrude** contextual tab. The part will be modified. Now, click on the assembly file name in the Model Tree and click on the **Activate** button from the Mini toolbar displayed; refer to Figure-9 to return to assembly environment.

*Figure-9. Activate button for assembly*

## Creating Part in Assembly

- Click on the **Create** button from the **Component** panel of the **Model** tab in the **Ribbon**. The **Create Component** dialog box will be displayed; refer to Figure-10.

*Figure-10. Create Component dialog box*

- Make sure the **Part** radio button is selected in the **Type** area and **Solid** radio button is selected in the **Sub-Type** area of the dialog box. Type the name as Lever in the **Name** edit box.
- Click on the **OK** button from the dialog box. The **Creation Options** dialog box will be displayed; refer to Figure-11.

*Figure-11. Creation Options dialog box*

## Using Copy From Existing option

- If you want to use any already created file as template then select the **Copy from existing** radio button from the **Creation Method** area of the dialog box and click on the **Browse** button to select desired file. The **Choose template** dialog box will be displayed.
- Select the desired template and click on the **OK** button. The selected file will get selected as template.
- Click on the **OK** button from the dialog box. The **Component Placement** contextual tab will be displayed and you will be asked to place the template part using assembly constraints.
- Place the part using desired constraints and click on the **OK** button from the contextual tab. Now, you can modify the part as discussed earlier.

## Using Locate Default Datums option

- Select the **Locate Default Datums** option from the **Creation Options** dialog box and select the desired method of locating datums from the **Locate Datums Method** area of the dialog box; refer to Figure-12.

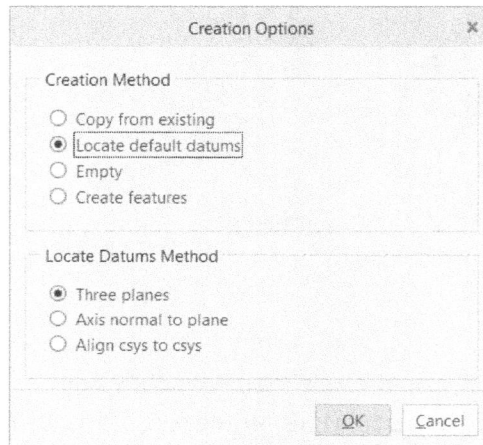

*Figure-12. Locate default datums option*

- Click on the **OK** button from the dialog box. You will be asked to select the reference according to radio button selected in the **Locate Datums Method** area of the dialog box.
- Select the datums and then create the model using the part modeling tools; refer to Figure-13.

*Figure-13. Creating part in assembly*

- Activate the assembly file from **Model Tree** as discussed earlier after creating part. Now, you can redefine the assembly constraints of the created part.

## Using Empty option

- Select the **Empty** radio button from the **Creation Options** dialog box and click on the **OK** button. A new empty part will be created. Modify the newly added part using the method discussed earlier. Note that on using this option, you will not be able to define assembly constraints for the component. So, you should not create part using this option if you are concerned about assembly constraints.

## Create Features option

- Select the **Create Features** radio button from the **Creation Options** dialog box and click on the **OK** button. A new part file will be created and the part modeling environment will be displayed; refer to Figure-14.
- Create the part using the modeling tools. Note that you can use the assembly geometry as reference while creating the part. For example, you can project the edges of shuttle part while creating sketch for feature or you can use the top face of shuttle as extrude depth reference; refer to Figure-15.

*Figure-14. Part modeling environment*

*Figure-15. Face of assembly part selected as reference*

Note that on using this option, you will not be able to define assembly constraints for the component. So, you should not create part using this option if you are concerned about assembly constraints.

## MECHANISM DESIGN

The tools for designing mechanism are available in the Mechanism application of Creo Parametric. The procedure to start Mechanism application is given next.

- Click on the **Mechanism** tool from the **Motion** panel of the **Applications** tab in the **Ribbon** while in Assembly environment; refer to Figure-16. The **Mechanism** tab will be added to the **Ribbon**; refer to Figure-17.

*Figure-16. Mechanism button*

*Figure-17. Mechanism tab in Ribbon*

Note that for using Mechanism tools, you must have the assembly constraints defined by Mechanism constraints like Pin, Cylinder, Rigid etc. Various tools in the **Mechanism** tab and their functions are discussed next.

## APPLYING GEAR RELATIONS

Gears are the backbone of modern locomotives. You can create real dynamics of a gear mechanism in the Mechanism environment of Creo Parametric. The procedure to do so is given next.

- Create an assembly of gears with Pin mechanism constraint applied to them; refer to Figure-18.

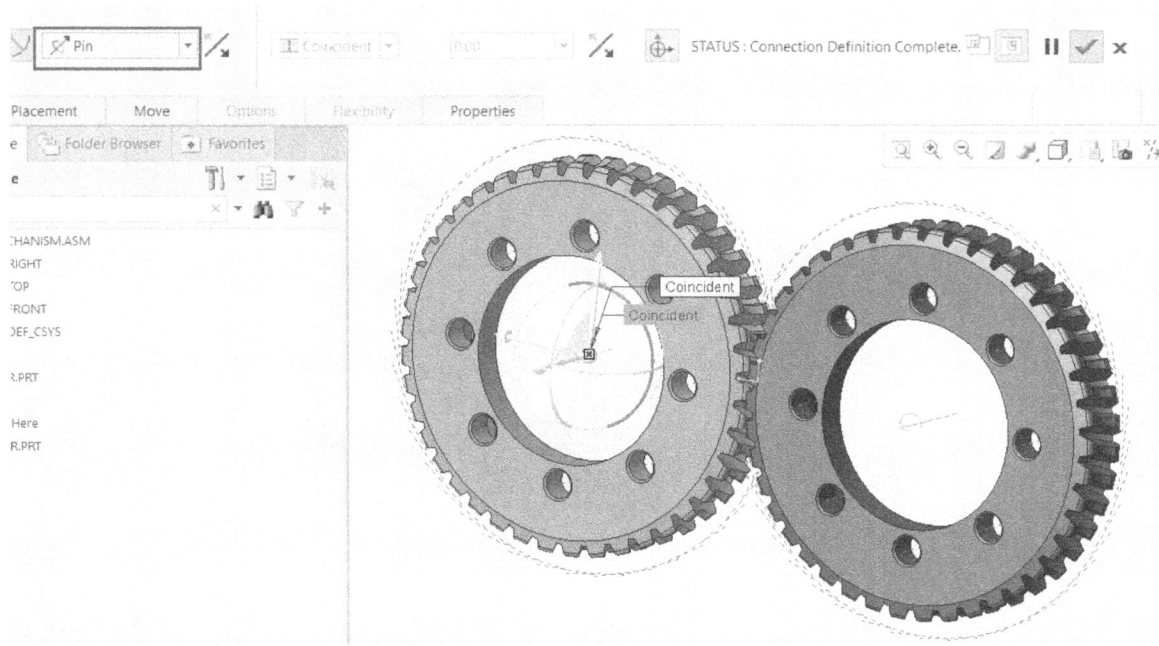

*Figure-18. Assembly of gears*

- Start the Mechanism application if not started yet. Click on the **Gears** tool from the **Connections** panel of the **Mechanism** tab of the **Ribbon**. The **Gear Pair Definition** dialog box will be displayed; refer to Figure-19. Also, you will be asked to select the motion axis.

*Figure-19. Gear Pair Definition dialog box*

- Select the motion axis of first gear. A double-arrow will be displayed on the motion axis; refer to Figure-20.

**Motion axis selected**

*Figure-20. Selecting motion axis*

- Specify the desired value of pitch circle diameter in the **Diameter** edit box of the **Pitch Circle** area of the dialog box; refer to Figure-21. Note that the gear ratio will be calculated based on the Pitch Circle Diameter specified here.

*Figure-21. Diameter edit box*

- Click on the **Gear 2** tab in the dialog box. You will be asked to select the motion axis of second gear. If you are not prompted to select the axis then click on the selection button ⬚ from the **Motion axis** area of the dialog box.
- Select the motion axis of the second gear. The motion arrow will be displayed on the motion axis. Note that this time the direction is opposite because mating gears rotate in opposite direction. If you want to reverse the direction of gear motion then click on the **Flip** button in **Motion axis** area of the dialog box.
- Specify the Pitch Circle Diameter for Gear 2 in the **Diameter** edit box.
- Click on the **Properties** tab to change the properties of the gear mechanism relation. The options in the dialog box will be displayed as shown in Figure-22.

*Figure-22. Properties options*

- Click in the drop-down of **Gear Ratio** area and select the **User defined** option to manually specify the gear ratio in edit boxes below the drop-down.
- Note that you have performed the whole process for Generic gears. If you want to create a spur gear relation then select the **Spur** option from the **Type** drop-down. The dialog box will be displayed as shown in Figure-23.

*Figure-23. Gear Pair Definition dialog box for spur gear*

- Specify the parameters related to spur gear like pressure angle and helix angle in respective edit boxes of **Properties** tab.
- In the same way, you can make Bevel gear, Worm gear, or Rack and Pinion mechanism relation by selecting the respective option from the **Type** drop-down.
- Click on the **OK** button to create the mechanism relation.
- Click on the **Drag Components** button and drag one of the gear to rotate it. The other gear will rotate accordingly. Save the file at desired location.

## APPLYING CAM-FOLLOWER RELATION

The most common example of cam-follower relation for a mechanical engineer is cam valve system in an engine. In Creo Parametric, we can create this relation by using the **Cams** tool. The procedure to create Cam-follower relation is given next.

- Create/open a cam-follower assembly; refer to Figure-24.
- Start Mechanism application if not started yet.

*Figure-24. Cam follower assembly*

- Click on the **Cams** tool from the **Connections** panel of the **Mechanism** tab in the **Ribbon**. The **Cam-Follower Connection Definition** dialog box will be displayed; refer to Figure-25 and you will be asked to select the surface/curve of cam.

*Figure-25. Cam-Follower Connection Definition dialog box*

- Select the surfaces/curves of Cam profile while holding the CTRL key or select the **Autoselect** check box in **Surfaces/Curves** area of the dialog box and then select one of the cam profile surface/curve, all the adjoining surfaces/curves will be selected automatically. After selecting the faces, press the MMB.

- Click on the Cam 2 tab from the dialog box. You will be asked to select the face of follower.
- Select the face of follower as shown in Figure-26 and press MMB.

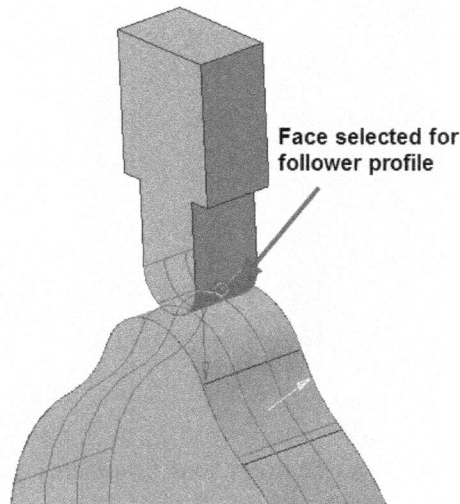

*Figure-26. Face selected for follower profile*

- Click on the **Properties** tab of the dialog box and specify the parameters like liftoff and friction coefficient in respective fields if required; refer to Figure-27.

*Figure-27. Properties tab of Cam-Follower Connection dialog box*

- Click on the **OK** button from the dialog box to create the connection.
- Click on the **Drag Components** button and drag the cam to rotate it. The follower will move up-down accordingly. Save the file at desired location.

## APPLYING 3D CONTACT RELATION

The 3D Contacts is used keep two bodies in contact with each other during motion. You can select references like sphere, cylinder, or plane to establish 3D contact. The procedure to apply this relation is given next.

- Create or open the assembly of geneva mechanism (You can find the assembly in resources of this book; refer to Figure-28.

*Figure-28. Geneva mechanism*

- Start the Mechanism application if not started yet. Now, using the **Drag Components** tool in the **Motion** panel, try to rotate the driver disk. You will find that pin does not rotate the driven disk but it passes through it. This is not desired in geneva mechanism, so we will apply 3D contact here to make software understand our purpose.
- Click on the **3D Contacts** tool from the **Connections** panel in the **Mechanism** tab of the **Ribbon**. The **3D Contact** contextual tab will be displayed in the Ribbon; refer to Figure-29 and you will be asked to select the contact references.

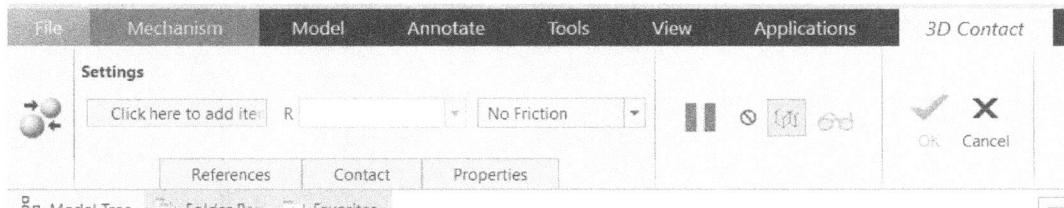

*Figure-29. 3D Contact contextual tab*

- Click on the **References** tab and select the round face of pin; refer to Figure-30. You will be asked to select face of second reference.
- Select the faces of driven disk as shown in Figure-31.

*Figure-30. Face of pin selected*

**Slot faces of driven disk selected**

*Figure-31. Faces of driven disk selected*

- Click on the **Contact** tab and specify the material properties for both sides of contact; refer to Figure-32.

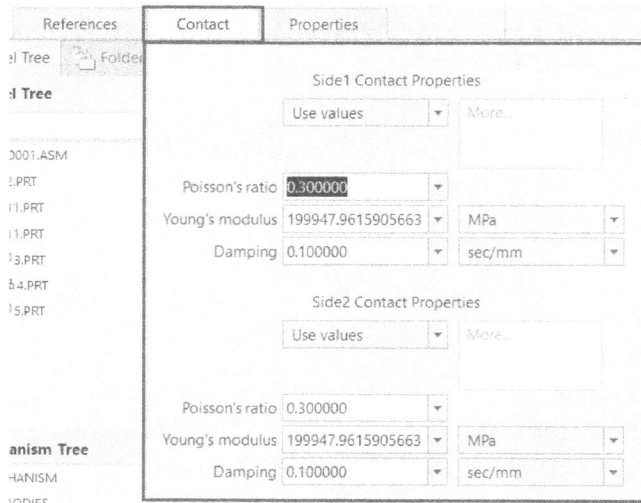

*Figure-32. Contact tab*

- If you want to select a material in place of explicitly specifying properties then select the **Select material** option from the drop-down in the **Contact** tab; refer to Figure-33 and click in the **More** field next to it. The **Materials** dialog box will be displayed as discussed in earlier chapters. Select the desired material from the dialog box and click on the **OK** button.

*Figure-33. Select material option*

- If you want to specify friction between the two contact references then select the **With Friction** option from the **Friction** drop-down of contextual tab in **Ribbon** and specify desired friction coefficients.
- Click on the **OK** button to create the relation. Save the file at desired location, we will run this mechanism later.

## APPLYING BELT RELATION

Belt mechanism has many uses in daily life and industries. You can find a belt in the car which runs the engine and alternator. You can also find belts in wheat crushing machines. The procedure to apply the belt relation is given next.

- Create or open the assembly in which you want to apply the belt relation; refer to Figure-34.

*Figure-34. Assembly for applying belt mechanism*

- Start the **Mechanism** application and click on the **Belts** tool from the **Connections** panel of the **Mechanism** tab in the **Ribbon**. The **Belt** contextual tab will be displayed; refer to Figure-35. Also, you will be asked to select the faces/edges of both pulleys.

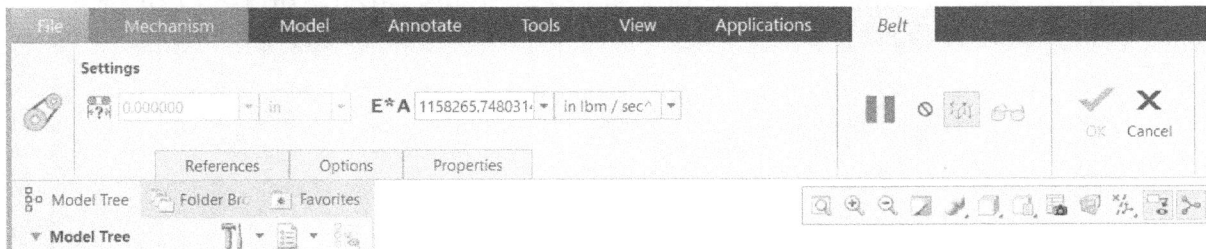

*Figure-35. Belt contextual tab*

- Select round face/edge of first pulley and then of second pulley while holding the CTRL key; refer to Figure-36. Preview of belt relation will be displayed; refer to Figure-37.

*Figure-36. Selecting edges of pulleys*

*Figure-37. Preview of belt relation*

- Click in the **Belt plane** selection box of **References** tab in the Belt contextual tab and select the desired plane for belt placement; refer to Figure-38.

*Figure-38. Belt plane selection*

- Click on the **Options** tab and specify the desired number of wraps of belt on each pulley; refer to Figure-39.

*Figure-39. Number of wraps option*

- Specify the tension of the belt in **E*Y** edit box of the **Ribbon**. Click on the **OK** button to create the belt relation. Save the file at desired location, we will analysis the motion later in this chapter.

# SERVO MOTORS

In Creo Mechanism application, there is always a need of motion maker that rotates/translates one or more objects in the mechanism. One of the motion making object in Mechanism is servo motor. Servo motors are like electrical motors which give rotary motion to the connected object. The procedure to apply servo motor is given next.

## Applying Rotational Motor

- Open the assembly file of geneva mechanism earlier created. Start the Mechanism application.
- Click on the **Servo Motors** tool from the **Insert** panel in the **Mechanism** tab of the **Ribbon**. The **Motor** contextual tab will be displayed; refer to Figure-40. Also, you will be asked to select the entity to be driven by motor.
- Select the rotational axis of driver disk in the assembly; refer to Figure-41. The **Rotational motion** button will get selected automatically.

*Figure-40. Motor contextual tab*

Rotation axis selected for motor

*Figure-41. Rotation axis selected for motor*

- Click on the **Profile Details** tab in the **Ribbon**. The options to modify motor profile will be displayed; refer to Figure-42.

*Figure-42. Profile Details tab*

- Click in the **Driven Quantity** drop-down and select desired parameter (like Velocity) to be used for applying motion. Specify the desired value of parameter in the **Coefficients** edit box.
- If you want to set a function for parameter value then click in the **Function Type** drop-down and select the desired function; refer to Figure-43. Specify the desired parameters for the function.

*Figure-43. Function Type drop-down*

- Click on the **OK** button from the **Ribbon** to apply motor. Save the file.

## Applying Translational Motor

Translational motor is useful when we need to check the effect of translational motion of a component in assembly. For example, in Rack and Pinion assembly; refer to Figure-44.

*Figure-44. Rack pinion assembly*

- Click on the **Servo Motors** tool from the **Insert** panel in the **Mechanism** tab of the **Ribbon**. The **Motor** contextual tab will be displayed as discussed earlier.
- Select the translational motion axis of model created by slider joints; refer to Figure-45.

Box like icon of translational motion axis selected

*Figure-45. Translational motion axis selected*

- Now, specify the position, velocity, or acceleration as discussed in previous topic.

Note that the part files with mechanism applied to them are available in the resources of this book.

## APPLY FORCE MOTORS

Force motors are used to represent the force/torque applied to a component in assembly. The procedure to apply force motor is given next.

- Click on the **Force Motors** tool from the **Insert** panel in the **Mechanism** tab of the **Ribbon**. The **Motor** contextual tab will be displayed in the **Ribbon**. Also, you will be asked to select the joint axis or component on which force motor is to be applied.
- Select the linear joint axis of assembly; refer to Figure-46. Click on the **Flip** button from the **Ribbon** to change the direction of force if required.

*Figure-46. Axis of joint to be selected*

- Click in the **Profile Details** tab of the **Ribbon** and specify the value of force in respective edit box.
- Set the other parameters as required and then click on the **OK** button.
- Save the file as we will later perform analysis on these mechanism.

In the same way, you can use the **Force/Torque** tool.

# APPLYING SPRINGS

The **Spring** tool is used to apply dynamic representation of spring in the model. The procedure to apply spring is given next.

- Click on the **Spring** tool from the **Insert** panel in the **Mechanism** tab of the **Ribbon**. The **Spring** contextual tab will be displayed; refer to Figure-47.

*Figure-47. Spring contextual tab*

- Select the desired button from the contextual tab to specify type of spring. Click on the **Extension and Compression Springs** button ⊣ to create an extension/compression spring or click on the **Torsion Springs** button ⋃ to create a torsion spring.
- Select two vertices of assembly if you are creating extension/compression spring; refer to Figure-48. Select the rotational joint axis of the assembly if you want to create torsion spring; refer to Figure-49.
- Specify the desired value of spring stiffness in the **Stiffness coefficient** edit box of the **Ribbon**. Select the desired unit of stiffness from the drop-down next to edit box.
- Set the desired equilibrium distance and click on the **OK** button.

*Figure-48. Vertices selected for extenstion spring*

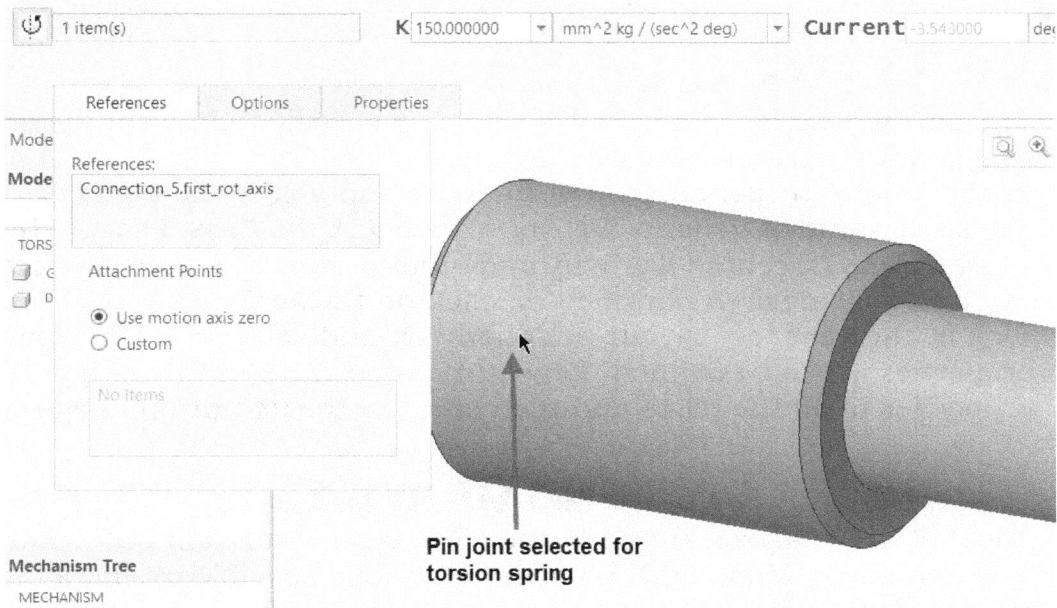

*Figure-49. Torsion spring*

The Dampers are applied in the same way as springs.

## SPECIFYING MASS PROPERTIES

Mass properties of a component define the behavior of that component due to its mass like it defines moment of inertia, effect of gravity on component and so on. The procedure to specify mass properties of assembly components is given next.

• Click on the **Mass Properties** tool from the **Properties and Conditions** panel of the **Mechanism** tab in the **Ribbon**. The **Mass Properties** dialog box will be displayed; refer to Figure-50.

*Figure-50. Mass Properties dialog box*

- Select the part whose mass properties are to be specified. Some of the options in the dialog box will become available to choose.
- Click in the **Define properties** by drop-down and select the **Density** option to specify desired value of density. Or, you can select the **Mass Properties** option and specify the mass, center of gravity and other parameters for the selected part.
- After specifying the desired parameters, click on the **Apply** button.
- Now, click in the **Part or Layout** selection box and select the other part whose mass properties are to be defined. Repeat the procedure until you have specified mass properties of all the parts and then click on the **OK** button to exit.

## DEFINING GRAVITY

Although most of the analyses are performed for objects on Earth but if you want to test your mechanism on Moon or Jupiter then it is also possible by changing gravity value!! The procedure to define gravity for mechanism is given next.

- Click on the **Gravity** tool from the **Properties and Conditions** panel of the **Mechanism** tab in the **Ribbon**. The **Gravity** dialog box will be displayed; refer to Figure-51.

*Figure-51. Gravity dialog box*

- Specify the desired magnitude of gravity in the **Magnitude** edit box. For Earth it is **9806.65**.
- Set the desired ratio for X, Y, and Z directions in their respective edit boxes to define direction of gravity force.
- After specifying the parameters, click on the **OK** button to apply gravity on mechanism.

## DEFINING INITIAL CONDITIONS

Initial condition is a very important aspect of dynamic analysis. In initial conditions we specify the starting position of component, initial velocity, motion axis velocity and so on. The procedure to specify initial condition is given next.

- Click on the **Initial Conditions** tool from the **Properties and Conditions** panel of the **Mechanism** tab in the **Ribbon**. The **Initial Condition Definition** dialog box will be displayed as shown in Figure-52.

*Figure-52. Initial Condition Definition dialog box*

- If you now click on the **OK** button then the current state of model will be saved as initial condition for mechanism which can later be used as initial state of assembly for analysis.
- If your model has some initial velocity then select the desired button from the left panel in the **Velocity conditions** area of the dialog box. The procedures to use each button in this area are discussed in next topic.
- After specifying velocity conditions, click on the **OK** button from the dialog box.

### Defining Initial Velocity of a Point

- Click on the **Define velocity of a point** button from the left panel in the **Velocity conditions** area of the dialog box. You will be asked to select the vertex whose velocity is to be defined.

- Click on the desired vertex of the model. The options in the dialog box will be displayed as shown in Figure-53.
- Specify the desired value of velocity in the **Magnitude** edit box.
- Set the required ratio of X, Y, and Z to define direction of velocity.
- Click on the **OK** button to apply velocity condition.

*Figure-53. Options for velocity*

## Defining Motion Axis Velocity

- Click on the **Define motion axis velocity** button ↳ from the left panel in the **Velocity conditions** area of the dialog box. You will be asked to select the motion axis.
- Select the motion axis and enter the velocity value in the **Magnitude** edit box of the dialog box.

## Defining Angular Velocity

- Click on the **Define angular velocity** button ↴ from the left panel in the **Velocity conditions** area of the dialog box. You will be asked to select the part to which angular velocity is being applied.
- Select the part. Preview arrow of angular velocity will be displayed.
- Specify the angular velocity in the **Magnitude** edit box in dialog box and also define the ratio of X, Y, and Z to specify direction of angular velocity.

## Defining Tangential Slot Velocity

- Click on the **Define tangential slot velocity** button ⥿ from the left panel in the **Velocity conditions** area of the dialog box. You will be asked to select the slot-follower connection to which velocity is being applied.

- Click on the slot-follower connection and enter the velocity value in **Magnitude** edit box of the dialog box; refer to Figure-54.

*Figure-54. Slot follower connection selected*

Note that you can define as many velocity conditions as required but they must form a valid relation. To check whether the initial conditions specified are valid, click on the **Evaluate model with velocity conditions** button from the left panel in the dialog box. To delete any velocity condition, select it from the list and then click on the **Delete highlighted condition** button ✕.

## DEFINING TERMINATION CONDITIONS

Termination condition is used to specify the conditions where the dynamic analysis should be stopped. To use termination conditions, we need some variables that represent dynamic features of assembly components. The procedure to create variables and apply termination conditions are discussed next.

## Creating Variables for Mechanism

- Click on the **Measures** tool from the **Analysis** panel in the **Mechanism** tab of the **Ribbon**. The **Measure Results** dialog box will be displayed; refer to Figure-55.

*Figure-55. Measure Results dialog box*

- Click on the **Create new measure** button ☐ from the **Measures** area of the dialog box. The **Measure Definition** dialog box will be displayed; refer to Figure-56 and you will be prompted to select a point or motion axis.

*Figure-56. Measure Definition dialog box*

- Click in the **Name** edit box of the dialog box and specify the desired name. (Like, we have specified "**Max Travel**".)
- Click in the drop-down of **Type** area and select the desired type of measurement variable. Short description of each measure type is given next. We will use Position type for this topic. You can use the other variable types accordingly.

- **Position** — Measure the location of a point, vertex, or motion axis during the analysis.
- **Velocity** — Measure the velocity of a point, vertex, or motion axis during the analysis.
- **Acceleration** — Measure the acceleration of a point, vertex, or motion axis during the analysis.
- **Connection reaction** — Measure the reaction forces and moments at joint, gear-pair, cam-follower, or slot-follower connections.
- **Net load** — Measure the magnitude of a force load on a spring, damper, servo motor, force, torque, or motion axis. You can also confirm the force load on a force motor.
- **Loadcell reaction** — Measure the load on a load cell lock during a force balance analysis.
- **Impact** — Determine whether impact occurred during an analysis at a joint limit, slot end, or between two cams.
- **Impulse** — Measure the change in momentum resulting from an impact event. You can measure impulses for joints with limits, for cam-follower connections with liftoff, or for slot-follower connections.
- **System** — Measure several quantities that describe the behavior of the entire system.
- **Body** — Measure several quantities that describe the behavior of a selected body.
- **Separation** — Measure the separation distance, separation speed, and change in separation speed between two selected points.
- **Cam** — Measure the curvature, pressure angle, and slip velocity for either of the cams in a cam-follower connection.

- **User defined** — Define a measure as a mathematical expression that includes measures, constants, arithmetical operators, parameters and algebraic functions.

- Select the **Position** option from the drop-down in **Type** area if not selected by default. You will be asked to select a vertex/motion axis to define position of assembly component.
- Click on the vertex of assembly component to whom a position variable is to be applied; refer to Figure-57.

*Figure-57. Vertex and coordinate system selected for position variable*

- Click on the selection button in the Coordinate System area of the dialog box and select the desired coordinate system; refer to Figure-57.
- Click in the drop-down of **Component** area in the dialog box and select the desired directional component of position to be measures. We have selected the **X-component** option as we want to measure horizontal distance covered by rack gear.
- From the **Evaluation Method** drop-down, select the desired option to specify evaluation method for measuring the selected component of position. Like, we are concerned about the Maximum distance covered by rack so we have selected **Maximum** option.
- Click on the **OK** button from the dialog box to create the variable. The variable will be created and you will return to **Measure Results** dialog box.
- Create more variables if required by using the same method. Click on the **Close** button from the **Measure Results** dialog box to exit.

## Specifying Termination Conditions

- Click on the **Termination Conditions** tool from the **Properties and Conditions** panel in the **Mechanism** tab of the **Ribbon**. The **Termination Condition Definition** dialog box will be displayed; refer to Figure-58.

*Figure-58. Termination Condition Definition dialog box*

- Specify the desired name for termination condition in the **Name** edit box.
- Click on the **Insert variable from list** button ⌐⌐ from the **Termination condition** area of the dialog box to use variables earlier saved.  The **Variables** dialog box will be displayed; refer to Figure-59.

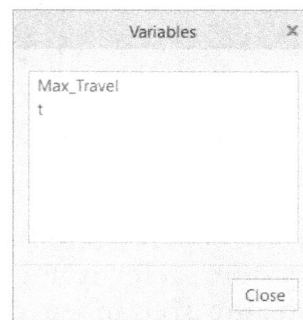

*Figure-59. Variables dialog box*

- Double-click on the variable in the dialog box to insert it in the expression and create the equation for termination condition; refer to Figure-60.

*Figure-60. Equation for termination condition*

- Click on the **OK** button after creating the equation to exit the dialog box.

Now, we are ready to perform analysis on the mechanism as we have applied motors/ force, initial conditions, termination conditions and other parameters of mechanism analysis. The procedure to perform analysis is given next.

# PERFORMING MECHANISM ANALYSIS

Before we move on to perform mechanism analysis, it is important to understand that in Creo Parametric Mechanism Analysis, we can perform five type of analyses: Position Analysis, Kinematic Analysis, Dynamic Analysis, Static Analysis, and Force balance analysis. The description of each analysis is given next.

## Position Analysis

A position analysis is a series of assembly analyses driven by servo motors. Only motion axes or geometric servo motors can be included in position analyses. Force motors do not appear in the list of possible motor selections when adding a motor for a position analysis.

- Position analyses were called kinematic or repeated assembly analysis in previous releases of Mechanism Design.
- If you edit an analysis that you created as a kinematic or repeated assembly analysis in a previous release of Mechanism Design, it will be regenerated as a position analysis.

A position analysis simulates the mechanism's motion, satisfying the requirements of your servo motor profiles and any joint, cam-follower, slot-follower, or gear-pair connections, and records position data for the mechanism's various components. It does not take force and mass into account when performing the analysis. Therefore, you do not have to specify mass properties for your mechanism. Dynamic entities in the model, such as springs, dampers, gravity, forces/torques, and force motors, do not affect a position analysis.

Use a position analysis to study:

- Positions of components over time
- Interference between components
- Trace curves of the mechanism's motion

## Kinematic Analysis

Kinematics is a branch of dynamics that deals with aspects of motion apart from consideration of mass and force. A kinematic analysis simulates the mechanism's motion, satisfying the requirements of your servo motor profiles and any joint, cam-follower, slot-follower, or gear-pair connection. A kinematic analysis does not take forces into account. Therefore, you cannot use force motors, and you do not have to specify mass properties for your mechanism. Dynamic entities in the model, such as springs, dampers, gravity, forces/torques, and force motors, do not affect a kinematic analysis.

If your servo motor has a noncontinuous profile, an attempt is made to make the profile continuous before running a kinematic analysis. If the profile is such that the software cannot make it continuous, the motor is not used for the analysis.

Use a kinematic analysis to obtain information on:
- Position, velocity, and acceleration of geometric entities and connections.
- Interference between components.

- Trace curves of the mechanism's motion.
- Motion envelopes that capture the mechanism's motion as a part.

## Dynamic Analysis

Dynamic analysis is a branch of mechanics that deals with forces and their relation primarily to the motion, but sometimes also to the equilibrium, of bodies. You can use a dynamic analysis to study the relationship between the forces acting on a body, the mass of the body, and the motion of the body.

Keep the following key points in mind when running a dynamic analysis:
- Motion axis-based servo motors are active for the duration of a dynamic analysis. For this reason the From and To times derived from the time domain for the analysis appear as the uneditable Start and End values.
- You can add both servo and force motors.
- If your servo or force motor has a noncontinuous profile, an attempt is made to make the profile continuous before running a dynamic analysis. If the profile cannot be made continuous, the motor is not used for the analysis.
- You can add forces or torques using the **External loads** tab.
- You can turn gravity and friction on or off.

You can evaluate the positions, velocities, accelerations, and reaction forces at the beginning of your dynamic analysis by specifying a zero time duration and running as usual. A suitable time interval for the calculations is determined automatically. If you graph measures from the analysis, the graph will contain only a single line.

## Static Analysis

Statics is the branch of mechanics that deals with forces acting on a body when it is at equilibrium. Use a static analysis to determine the state of a mechanism when it is subject to known forces. The application searches for a configuration in which all the loads and forces in your mechanism balance and the potential energy is zero. A static analysis can identify a static configuration faster than a dynamic analysis can because it does not consider velocity in the calculation.

Keep the following key points in mind when running a static analysis:

- If you do not specify an initial configuration, the static analysis starts from the currently displayed position of the model when you click **Run**.
- When you run a static analysis, a graph of acceleration versus iteration number appears, showing the maximum acceleration of the mechanism's entities. As the analysis calculation proceeds, both the graph display and the model display change to reflect the intermediate positions reached during the calculation. When the maximum acceleration for the mechanism reaches 0, your mechanism has reached a static configuration.
- You can adjust the maximum step size between each iteration of the static analysis by changing the **Maximum step factor** on the **Preferences** tab of the **Analysis Definition** dialog box. Reducing this value reduces the positional change between each iteration and can be useful when analyzing mechanisms incorporating large accelerations.
- If a static configuration for your mechanism cannot be found, the analysis ends

and the mechanism remains in the last configuration reached during the analysis.
- Any measures computed will be for the final times and positions, not a time history for the settling process.

## Force Balance Analysis

A force balance analysis is an inverse static analysis. In a force balance analysis, you derive the resulting reaction forces from a specific static configuration, whereas, in a static analysis, you apply forces to a mechanism to derive the resulting static configuration. Use a force balance analysis to determine the forces required to keep a mechanism fixed in a particular configuration.

Before you can run a force balance analysis, you must reduce the number of degrees of freedom in the mechanism to zero. Use connection locks, body locks between two bodies, a load cell lock at a point, or apply active servo motors to motion axes. Use the items in the Analysis Definition dialog box to evaluate the DOFs on your mechanism and apply constraints to it until you achieve zero DOF.

The procedure of performing all these analysis is similar. Here, we will discuss the procedure of performing dynamic analysis. You can apply the procedure to other analysis in the same way.

## Performing Dynamic Analysis

- Click on the **Mechanism Analysis** button from the **Analysis** panel in the **Mechanism** tab of the **Ribbon**. The **Analysis Definition** dialog box will be displayed; refer to Figure-61.

*Figure-61. Analysis Definition dialog box*

- Specify the desired name of analysis in the **Name** edit box.
- Click in the **Type** drop-down and select the **Dynamic** option.
- Specify the parameters in **Preferences** tab like duration of analysis, frame rate for analysis, minimum interval in their respective edit box. If you want to evaluate the positions, velocities, accelerations, and reaction forces of the entities in your mechanism at the beginning of your dynamic analysis, enter 0 for Duration. You can use this method as a quick check before running a longer dynamic analysis. Note that **Frame Rate = 1/Interval & Frame Count = Frame Rate * Length + 1**.
- If you want lock any connection then click on the **Create connection lock** button from the **Locked entities** area of the dialog box. You will be asked to select the connection. Select a connection and press the MMB. The locked connection will be listed in the **Locked entities** area of the dialog box. Clear the check box for connection from the list if you want to deactivate the lock. In the same way, you can use other buttons of the area.
- Click on the **I.C.State** radio button from the **Initial configuration** area of the dialog box and select the desired initial condition from the drop-down. Similarly, select the desired termination condition from the **Termination condition** area of the dialog box.
- Click on the **Motors** tab and select the desired motors to be used for analysis.
- Click on the **External Loads** tab and select the forces/torques to be applied in mechanism.
- If you want to check the effect of friction and gravity in your analysis then make sure the **Enable gravity** and **Enable all friction** check boxes are selected in the **External Loads** tab of the dialog box.
- Click on the **Run** button from the dialog box to run the analysis. If the parameters specified are correct then the analysis will run as expected otherwise you need to recheck all the parameters.
- Click on the **OK** button from the dialog box. The analysis will be saved in the **Mechanism Tree**.

Using the **Playback** tool in **Analysis** panel in **Ribbon**, you can play the mechanism animation again. Now, you can run mechanism on all the examples of assemblies that we have discussed earlier. Part files for all the examples are available in resources.

# SELF ASSESSMENT

1. Write down the difference between Top-Down assembly approach and Bottom-Up assembly approach.

2. Discuss the procedure of performing a kinematic analysis using a suitable example in Creo Mechanism.

# FOR STUDENT NOTES

# Chapter 11

# Sheetmetal Design

## Topics Covered

The major topics covered in this chapter are:

- *Introduction*
- *Starting Sheetmetal*
- *Modeling tools in Sheetmetal environment*
- *Walls*
- *Bend, Unbend, Bend Back and Flat Pattern*
- *Part to Sheetmetal*
- *Rip tools*
- *Forming Tools*
- *Flexible Modeling for Sheetmetal*

Model

Flat Pattern

# INTRODUCTION

Sheet metal work is an important aspect of Mechanical engineering. Many parts around us are manufactured via sheetmetal processes. For example, car body, vents in houses, Air-conditioner ducts, spoon, metal bowls and so on. The sheetmetal parts generally have thickness ranging from fraction of millimeter to 12.5 millimeters i.e. up to half inch. Like welding and machining, sheetmetal also has its own processes like, bending, punching, stamping, spinning, rolling and so on. In this chapter, we will discuss about the tools available in Creo Parametric related to sheetmetal designing.

# STARTING SHEETMETAL DESIGN

*   Start Creo Parametric and click on the **New** button from the **Quick Access Toolbar**. The **New** dialog box will be displayed.
*   Click on the **Part** radio button in the **Type** area and **Sheetmetal** radio button from the **Sub-type** area of the dialog box.
*   Clear the **Use default template** check box and then click on the **OK** button from the dialog box. The **New File Options** dialog box will be displayed; refer to Figure-1.

*Figure-1. New File Options dialog box*

*   Select the **mmns part sheetmetal** option from the dialog box and then click on the **OK** button from the dialog box. The Sheetmetal environment of Creo Parametric will be displayed; refer to Figure-2.

*Figure-2. Sheetmetal environment*

There is a long list of tools used for creating sheetmetal parts in various panels of Creo Parametric. You will also find some familiar tools like Extrude, Revolve, and so on in these panels. In this chapter, we will be discussing sheetmetal application of all these familiar tools. We will also be discussing the tools that are specifically meant for sheetmetal. These tools are discussed next.

# EXTRUDE FOR SHEETMETAL

The **Extrude** tool in sheetmetal environment works similar to the one in Part environment but, in sheetmetal environment, you need to specify the thickness of sheet while extruding the sketch. The procedure is given next.

- Click on the **Extrude** tool from the **Shapes** panel in the **Ribbon**. You are asked to select a sketch or sketching plane.
- Select a sketch if you have already created otherwise, select a sketching plane.
- On selecting the sketching plane, sketching environment will be displayed. Create a sketch and click on the **OK** button from the **Ribbon**. Note that you can create close or open both type of sketches but the sketch must not be self-intersecting.
- On doing so, the preview of sheetmetal extrusion will be displayed; refer to Figure-3.

*Figure-3. Preview of sheetmetal extrusion*

- Specify the desired sheet thickness and click on the **OK** button to create the sheetmetal extrusion.

In the same way, you can use the other tools that have been discussed in chapter for Part Modeling; refer to Figure-4.

*Figure-4. Tools already discussed*

# TWIST

The **Twist** tool is used to create twisted sheetmetal parts. This tool can only be used after creating the base wall. Note that the first wall created in sheetmetal is called base wall. The base wall can be created by all the part modeling tools. The procedure to use **Twist** tool is given next.

- Click on the **Twist** tool from the expanded **Shapes** panel in the **Ribbon**. The **Twist** contextual tab will be displayed; refer to Figure-5 and you will be asked to select an edge to which twisted wall would be attached.

Figure-5. Twist contextual tab

- Select an edge of the base wall; refer to Figure-6. Note that this edge will become base sketch for twist feature.

Figure-6. Edge selected for twisting

- On selecting the edge, preview of the twisted wall will be displayed; refer to Figure-7.

Figure-7. Preview of twisted wall

- Set the desired length and angle of twisted wall by double-clicking on the respective dimensions in the model area or you can specify the value in edit boxes in the contextual tab.

- By default, the wall width of twisted wall is calculated using offset dimension of selected edge. If you want to define wall width by specifying desired value then click on the **Wall width at edge** button ⊞ and specify desired value in the next edit box; refer to Figure-8.

*Figure-8. Specifying width of twist feature at edge*

- By default, the twist axis is at the center of the selected edge. If you want to specify the location of twist axis explicitly then click on the **Set a datum point as the twist axis location** button ⊞ and click on the datum point through which the twist axis should pass; refer to Figure-9.

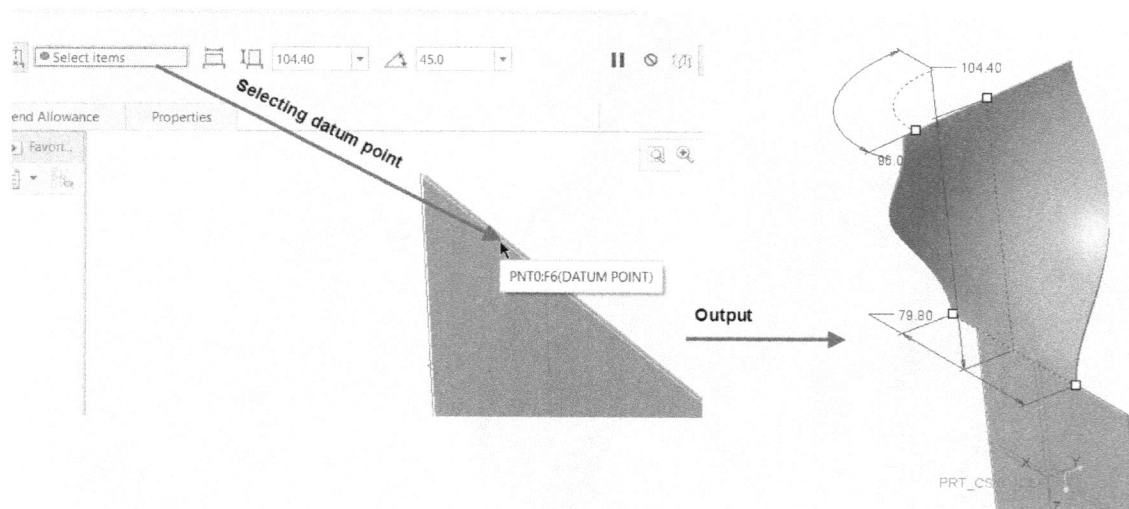

*Figure-9. Selecting datum point for axis*

- If you want to modify the width of twisted wall at the end then click on the **Allows end wall width modification** button ⊞ from the contextual tab and specify the desired value of width in next edit box.
- Click on the **OK** button from the contextual tab to create the twisted wall.

## FLAT TOOL

The **Flat** tool is used to create flat walls of sheetmetal. In manufacturing, these walls are created by bending or drawing processes. The procedure to use this tool is given next.

- Click on the **Flat** tool from the **Shapes** panel in the **Ribbon**. The **Flat** contextual tab will be displayed; refer to Figure-10 and you will be asked to select an edge.

*Figure-10. Flat contextual tab*

- Select an edge. Preview of the flat wall will be displayed; refer to Figure-11.

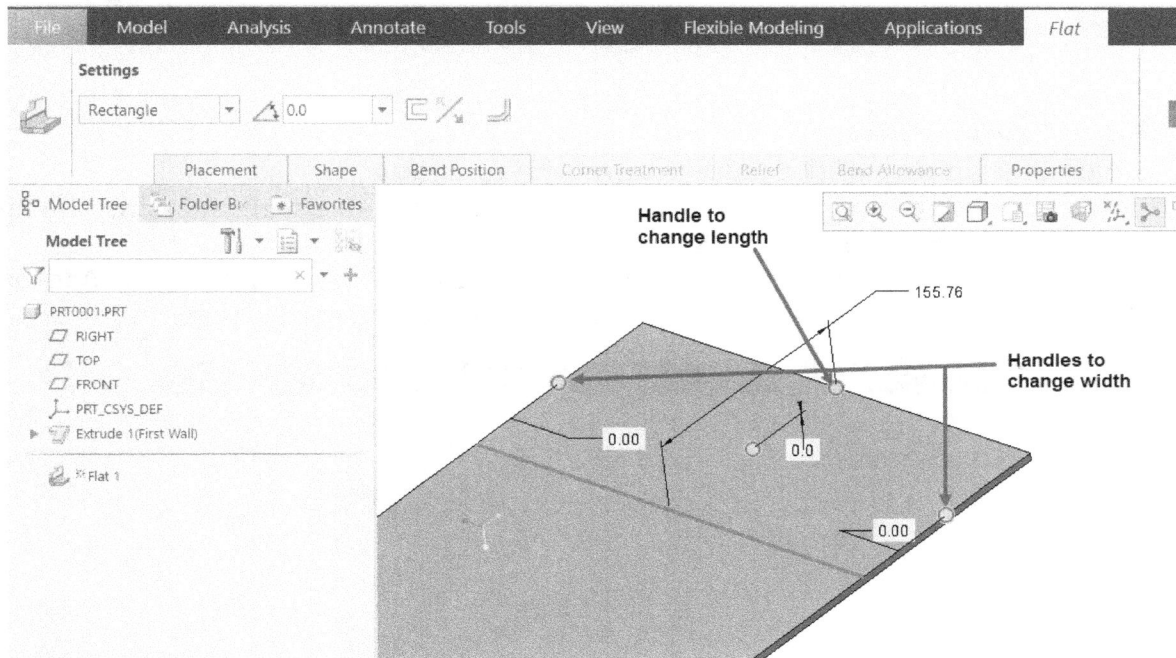

*Figure-11. Preview of flat wall*

- Using the handles displayed in preview, you can change the width, length and angle of the flat wall.
- Select the desired shape of wall from the **Shapes** drop-down in the **Ribbon**; refer to Figure-12. Preview of the wall will change accordingly; refer to Figure-13.

*Figure-12. Shapes drop-down*

Figure-13. Flat wall preview

- Double-click on any of the value in preview to change it.
- You can specify the angle of wall with respect to the base wall by using the **Angle** edit box in the contextual tab (**Ribbon**); refer to Figure-14.
- If you want to specify the bend radius at turning edge then click on the **Add bends** button from the contextual tab and specify the bend radius; refer to Figure-15.

Figure-14. Angle of wall specified

*Figure-15. Bend radius*

- Using the options in the **Shape** tab in the contextual tab, you can create desired shape of flat wall with the help of creating/created sketch; refer to Figure-16. Using the options in the **Shape** tab, you can create sketch for wall, use an already created sketch, and save the current shape of wall as a sketch by using the **Sketch**, **Open** and **Save as** buttons respectively.

*Figure-16. Shape tab*

- Click on the **Bend Allowance** tab to specify bending allowance for sheet; refer to Figure-17. You can specify K factor or Y factor, or you can use the bend table to specify bending allowance.

*Figure-17. Bend Allowance tab*

The K-Factor in sheet metal work is the ratio of the neutral axis to the material thickness. The K-Factor is used to calculate flat patterns because it is directly related to how much material is stretched during the bend.

The Y-Factor is simply a variable based on the more commonly used K-Factor. It is derived by taking half of the K-Factor multiplied by PI.

- After specifying the desired parameters, click on the **OK** button from the contextual tab to create the flat wall.

## FLANGE TOOL

The **Flange** tool works in similar way to **Flat** tool but in case of **Flange** tool, we can modify cross-section of wall in place of base sketch. The procedure to use this tool is given next.

- Click on the **Flange** tool from the **Shape** panel in the **Ribbon**. The **Flange** contextual tab will be displayed; refer to Figure-18.

*Figure-18. Flange contextual tab*

- Select an edge to create a flange wall attached to it. Preview of the flange wall will be displayed; refer to Figure-19.

*Figure-19. Edge selected for flange wall*

- To make multiple selections at a time, click on the **Placement** tab and then click on the **Details** button; refer to Figure-20. The **Chain** dialog box will be displayed; refer to Figure-21.

*Figure-20. Details button*

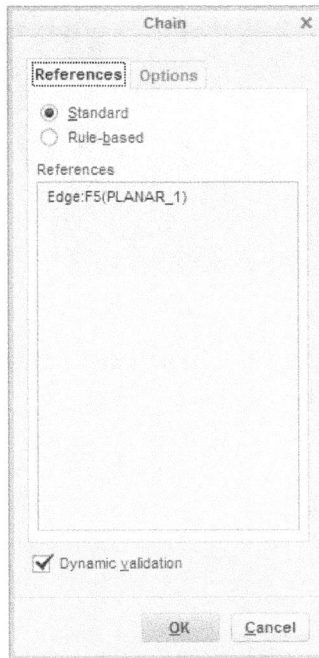

*Figure-21. Chain dialog box*

- Hold the **CTRL** key from Keyboard and select the edges one by one; refer to Figure-22.

*Figure-22. Multiple edges selected*

- Click on the **OK** button from the dialog box. Preview of the flange wall with multiple edges will be displayed; refer to Figure-23.

*Figure-23. Preview of flange wall*

- Using the options in the **Shape** tab, you can create desired cross-section for the flange wall; refer to Figure-24.

*Figure-24. Sketched cross-section of flange wall*

- Using the options in **Length** tab, you can specify the starting and end limit of the flange wall; refer to Figure-25.

*Figure-25. Flange wall after specifying desired length*

- Select the desired option from the Bend Position tab to define location of bend; refer to Figure-26.

*Figure-26. Defining Bend position*

- The options in the **Corner Treatment** tab are used to modify the junctions of edges. There are five options available in the drop-down to treat the corner junctions; refer to Figure-27. Preview of junctions with main options is given in Figure-28, Figure-29, Figure-30, and Figure-31.

*Figure-27. Corner Treatment options*

Figure-28. Open corner treatment

Figure-29. Overlap corner treatment

*Figure-30. Blind corner treatment*

*Figure-31. Gap corner treatment*

- The options in the **Miter Cuts** tab make their use in **Flat** pattern of sheetmetal part. Note that after creating the sheetmetal part, our final step is to create a flat pattern so that a sheet of that size can be cut and transformed into desired part. We will see the effect of miter cut while discussing flat patterns in this chapter.

- Click on the **Relief** tab to provide relied in sheetmetal part so that the part can be bent or unbent on the machine without material overlapping and loss. There are two categories by which we can apply relief; Bend relief and Corner relief; refer to Figure-32. The Bend relief is applied to set the shape of relief at the bends whereas Corner relief change the shape of relief at corners formed by walls; refer to Figure-33 and Figure-34.

*Figure-32. Relief categories*

*Figure-33. Bend relief preview*

*Figure-34. Corner relief*

- Select the desired options and specify related values to create desired relief in part. Note that you can specify Bend relief only when you bend the sheetmetal at some distance from edge ends. If the sheet is bent for the full edge then effect of Bend relief will not be displayed.
- Other options in the contextual tab are same as discussed for Flat wall earlier. After specifying the desired parameters, click on the **OK** button from the contextual tab to create the flange walls.

# PLANAR TOOL

The **Planar** tool is used to create base wall like extrude and other part modeling software. Using this tool, you can create base wall on the basis of a selected/created closed sketch. The procedure to use this tool is given next.

- Click on the **Planar** tool from the **Shapes** panel in the **Ribbon**. The **Planar** contextual tab will be displayed; refer to Figure-35 and you will be asked to select or create a sketch.

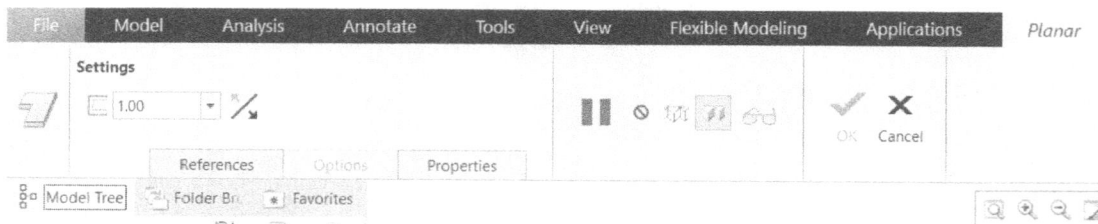

*Figure-35. Planar contextual tab*

- If you already have a closed sketch in the modeling area then select it otherwise, click on the **References** tab and click on the **Define** button to create a sketch.
- Create a closed sketch at desired plane and then click on the **OK** button to exit the sketching environment. Preview of the wall will be displayed; refer to Figure-36.

*Figure-36. Preview of planar wall*

- Click in the **Sheet Thickness** edit box and specify the desired thickness for the planar wall; refer to Figure-37.

*Figure-37. Sheet Thickness edit box*

- Click on the **OK** button to create the wall.

## BENDING TOOLS

The Bending tools, as the name suggests, are used to bend the sheet along the specified bend line. There are three tools in Creo Parametric to perform bending; Bend, Edge Bend, and Planar Bend; refer to Figure-38. These tools are discussed next.

*Figure-38. Bending tools*

## Bend tool

The **Bend** tool is used to bend a flat sheet using a bend line lying on its face. The procedure is given next.

- Click on the **Bend** tool from the **Bend** drop-down in the **Bends** panel of the **Ribbon**. The **Bend** contextual tab will be displayed; refer to Figure-39.

*Figure-39. Bend contextual tab*

- Click on the **Placement** tab. You will be asked to select a surface for defining bend line; refer to Figure-40.

*Figure-40. Placement tab of Bend contextual tab*

- Select a face of wall that you want to bend; refer to Figure-41. Preview of the bend line will be displayed.

*Figure-41. Face selected*

- One by one, drag the green handles of bend line to the references edges; refer to Figure-42.

*Figure-42. Dragging reference handle*

- If you face problem in dragging the handle, then you can define the references in the **Bend line** tab of contextual tab; refer to Figure-43.

*Figure-43. Bend line references*

- You can also use the **Sketch** button to sketch the bend line in place of selecting references. **Note that while sketching the bend line, you are allowed to create only one line**; refer to Figure-44.

Bend line sketch created

Preview of bending

Output of bending

*Figure-44. Bending by sketched line*

- The three buttons at the left most of the contextual tab are used to decide the bending side of material with respect to the bending line; refer to Figure-45. The effect of selecting a button from the three is given in Figure-46, Figure-47, and Figure-48.

*Figure–45. Bending side
with respect to bending line*

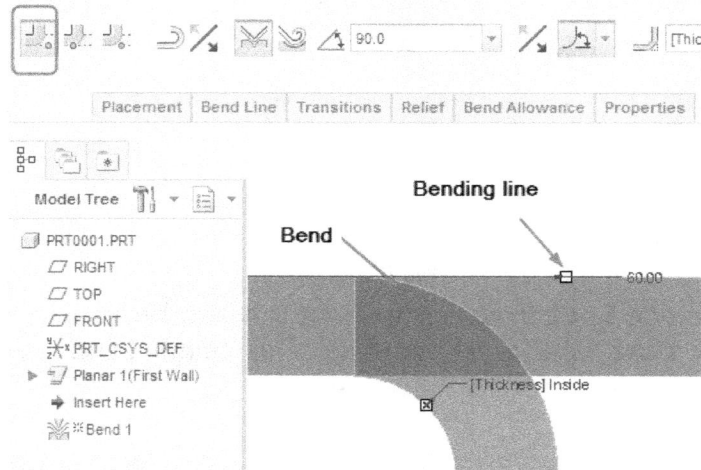

*Figure–46. Bend upto bending line*

*Figure–47. On other side of bend line*

*Figure-48. On both side of bend line*

- You can change the fixed side of bend by using the **Change location of fixed side** button.
- Till this point, we have used **Bend by angle** button. Which practically means, the sheet is bent on a punch press. But, we can also bend the sheet on a roller. To do so, click on the **Bend to end surface** button. The angle edit box in the **Ribbon** will be deactivated and you are asked to specify radius of bend.
- Specify the desired bend radius. Preview of the bend will be displayed; refer to Figure-49. Note that the preview will be displayed only when you have specified a possible radius in the edit box.

*Figure-49. Bend using bend to end surface button*

- Use the **Radius side** drop-down to change the radius from inner to outer and vice-versa.
- To apply the transition, click on the **Transition** tab and click on the **Add Transition** option. The **Sketch** button will become active. Click on it and draw two lines defining limits of transition; refer to Figure-50.

*Figure-50. Sketch of transition lines*

- Click on the **OK** button from the sketching environment. Preview of the bend with transition will be displayed; refer to Figure-51.

*Figure-51. Transition area of bend*

- After specifying the desired parameters, click on the **OK** button from the contextual tab to create the bend.

## Edge Bend

The **Edge Bend** tool is used to apply bend at the sharp edges of the sheetmetal part. The procedure is given next.

- Click on the **Edge Bend** tool from the **Bend** drop-down in the **Bends** panel of the **Ribbon**. The **Edge Bend** contextual tab will be displayed; refer to Figure-52.

*Figure-52. Edge Bend contextual tab*

- Select the sharp edges of the model. Preview of the bend will be displayed; refer to Figure-53.

*Figure-53. Edges selected*

- Specify the desired radius in the edit box and click on the **OK** button from the contextual tab to apply the bend. Note that you can provide relief and bend allowance to the bends in the same way as discussed earlier using the **Relief** and **Bend Allowance** tabs in the contextual tab.

## Planar Bend

The **Planar Bend** tool is used to bend the sheet by specified radius or angle in its own plane. The procedure is given next.

- Click on the **Planar Bend** tool from the **Bend** drop-down in the **Bends** panel of the **Ribbon**. The **Menu Manager** with two options will be displayed; refer to Figure-54.

*Figure-54. Menu Manager*

## Planar Bend using Angle

- Select the **Angle** option from the **Menu Manager** and click on the **Done** button (or press MMB). The **Bend Options** dialog box will be displayed with **Angle** options and **Use Table** options will be displayed in the **Menu Manager**.
- Select the **Part Bend Tbl** option and press the Middle button of Mouse (MMB). You are asked to select a sketching plane for specifying bending line.
- Select a plane. You are asked to specify sketching direction. Press the MMB and specify references as we do for sketching.
- Create a sketch line for specifying bend line; refer to Figure-55.

*Figure-55. Bending line*

- Click on the **OK** button from **Ribbon** to exit sketching environment.
- Select the **Flip** button from the **Menu Manager** displayed if you want to flip the bending side.
- Click **Okay** from the **Menu Manager**. You are asked to specify the direction of **Angle**.
- Click on the **Okay** button after specifying desired direction. You are asked to specify the angle; refer to Figure-56.

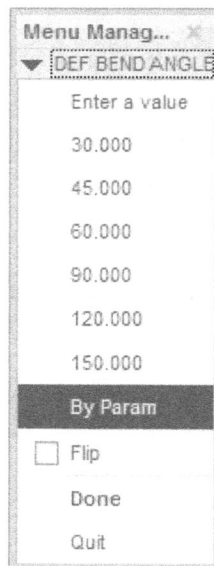

*Figure-56. Menu Manager with angle values*

- Specify the desired angle and click on the **Done** button from the **Menu Manager**. The **SEL RADIUS** option will be displayed in the **Menu Manager**. If you want to specify a custom radius then click on the **Enter Value** option and specify the radius otherwise, default parameters will be applied.

- After specifying radius, press the MMB. You are asked to specify direction of bend.
- Select desired side by using the **Flip** option and press MMB.
- Click on the **Preview** button from the dialog box; refer to Figure-57. If you want to make changes then select the desired option from the dialog box and click on the **Define** button otherwise, click on the **OK** button to create the bend.

*Figure-57. Preview of planar bend*

## Planar Bend using Roll

- Select the **Roll** option after clicking on the **Planar Bend** tool and then click on the **Done** option from the **Menu Manager**.
- Select the **Part Bend Tbl** option and press the Middle button of Mouse (MMB). You are asked to select a sketching plane for specifying bending line.
- Select a plane. You are asked to specify sketching direction. Press the MMB and specify references as we do for sketching.
- Create a sketch line for specifying bend line.
- Click on the **OK** button from **Ribbon** to exit sketching environment.
- Select the **Flip** button from the **Menu Manager** displayed if you want to flip the bending side.
- Click **Okay** from the **Menu Manager**. The **SEL RADIUS** option will be displayed in the **Menu Manager**. Click on the **Enter Value** option and specify the radius otherwise, default parameters will be applied.
- After specifying radius, press the MMB. You are asked to specify direction of bend.
- Select desired side by using the **Flip** option and press MMB.
- Click on the **Preview** button from the dialog box. Preview of the roll planar bend will be displayed; refer to Figure-58. If you want to make changes then select the desired option from the dialog box and click on the **Define** button otherwise, click on the **OK** button to create the bend.

*Figure-58. Preview of planar roll bend*

# UNBEND

The **Unbend** tools are used to perform the reversal of **Bend** tools. There are three tools to perform unbend operation. The most used of the three is **Unbend** tool which is discussed next.

## Unbend tool

• Click on the **Unbend** tool from the **Unbend** drop-down in the **Bends** panel of the **Ribbon**; refer to Figure-59. The **Unbend** contextual tab will be displayed with the preview of unbending in modeling area; refer to Figure-60.

*Figure-59. Unbend drop-down*

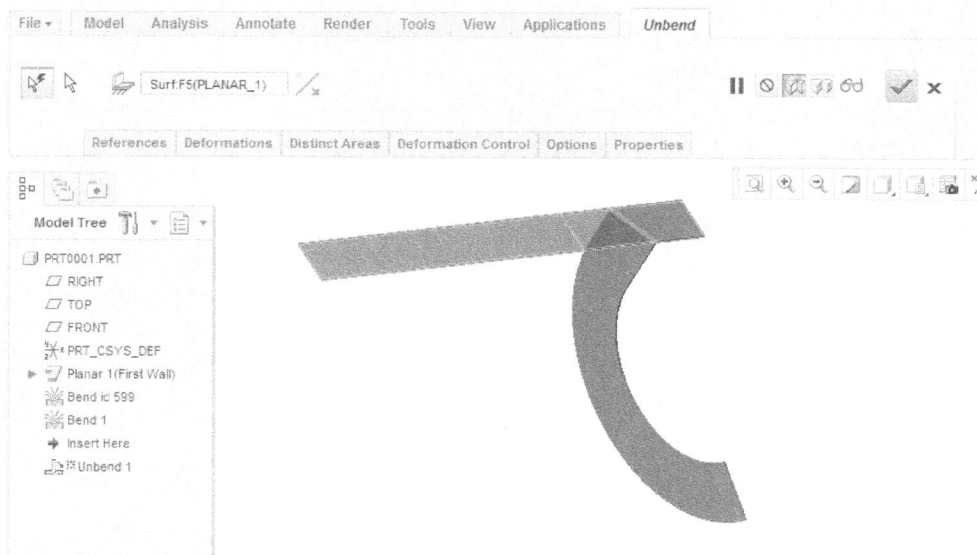

*Figure-60. Unbend contextual tab with unbending preview*

• Click on the **OK** button from the contextual tab to perform unbending.

Most of the time the Unbending process is simple but what to do if we have a closed sheetmetal part as shown in Figure-61.

*Figure-61. Closed sheetmetal part*

In such cases, we can used the **Deformation** options in the **Unbend** contextual tab. The steps are given next.

• After selecting the **Unbend** tool, click on the **Deformations** tab; refer to Figure-62.

*Figure-62. Deformations tab*

• Click in the **Deformation Surfaces collector** in the tab and select a face on which you want to apply the deformations so that our part can unbend; refer to Figure-63. Preview of deformation will be displayed; refer to Figure-64.

*Figure-63. Deformation surfaces collector*

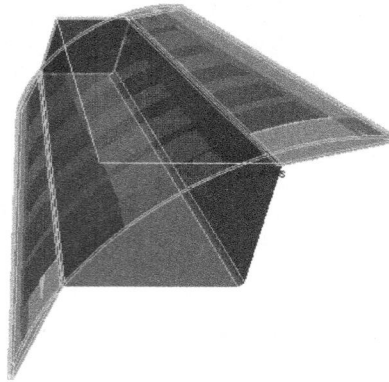

*Figure-64. Preview of deformation*

- But, still we can not unbend the part. So, click on the **Deformation Control** tab and select the **Rip area** radio button. Preview of unbent sheet will be displayed; refer to Figure-65.

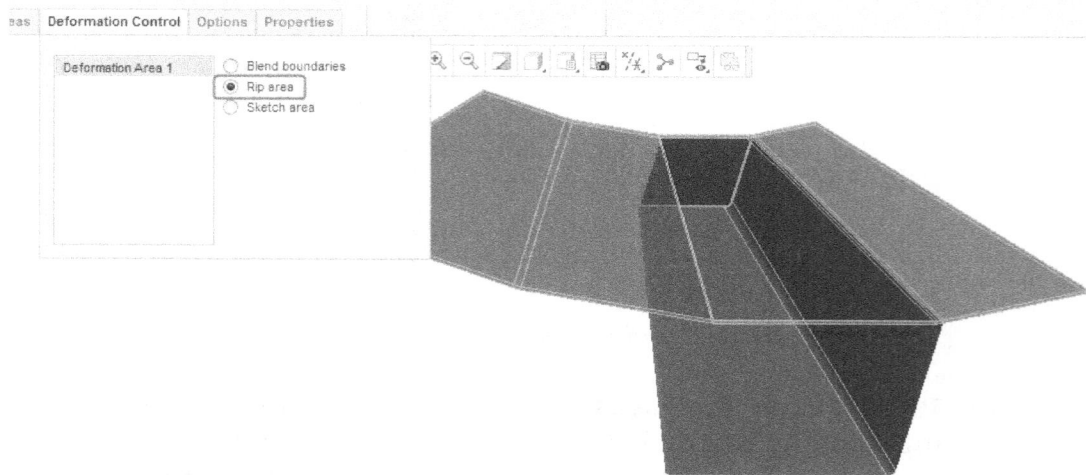

*Figure-65. Preview of unbent sheet*

- Click on the **OK** button to apply unbending.

# BEND BACK

The **Bend Back** tool is used to reverse the operation done by **Unbend** tool but it does not reverse the deformation done in the part during unbending. The procedure is given next.

- Click on the **Bend Back** tool from the **Bends** panel of the **Ribbon**. Preview of the bending back will be displayed with the **Bend Back** contextual tab; refer to Figure-66.
- Select the desired face to make it reference and click on the **OK** button from the contextual tab to create the feature.
- You can use the **Manual selection** button from the contextual tab and select each face to bent back manually; refer to Figure-67.

Figure-66. Bend Back contextual tab

Figure-67. Bend back preview for manual face selection

# BEND ORDER

The **Bend Order** tool is available in the expanded **Bends** panel of **Ribbon**. This tool is used to mark the order in which sheet metal need to be bent. The procedure to use this tool is given next.

• Click on the **Bend Order** tool from the expanded **Bends** panel in the **Model** tab of **Ribbon**. The **Bend Order** dialog box will be displayed with bend lines in the model; refer to Figure-68.

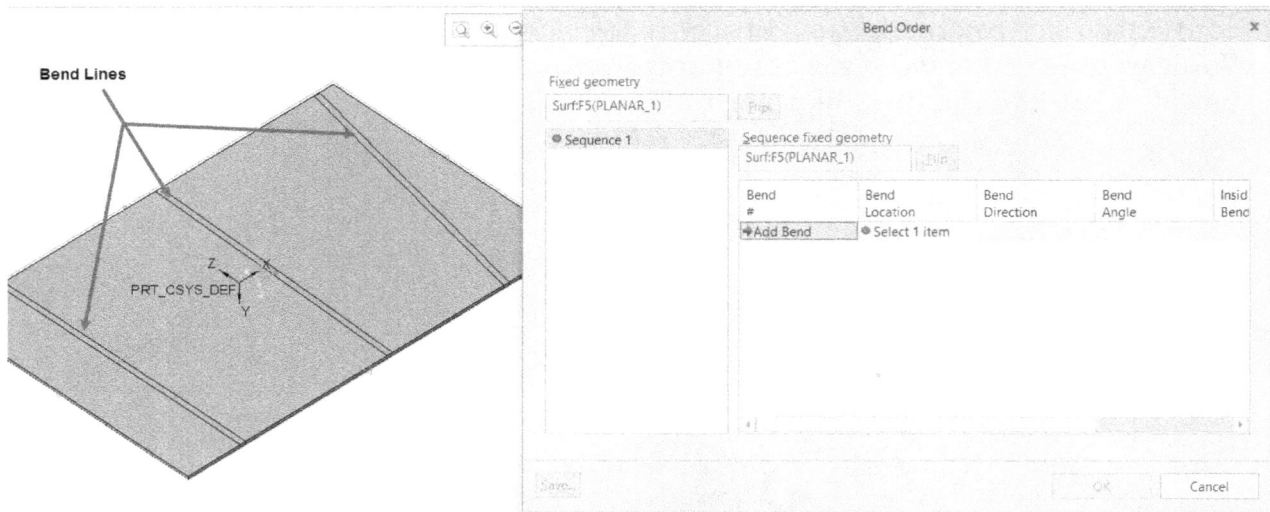

*Figure-68. Bend Order dialog box*

• Select the bend lines from the model in order by which you want to number bends in the flat pattern; refer to Figure-69.

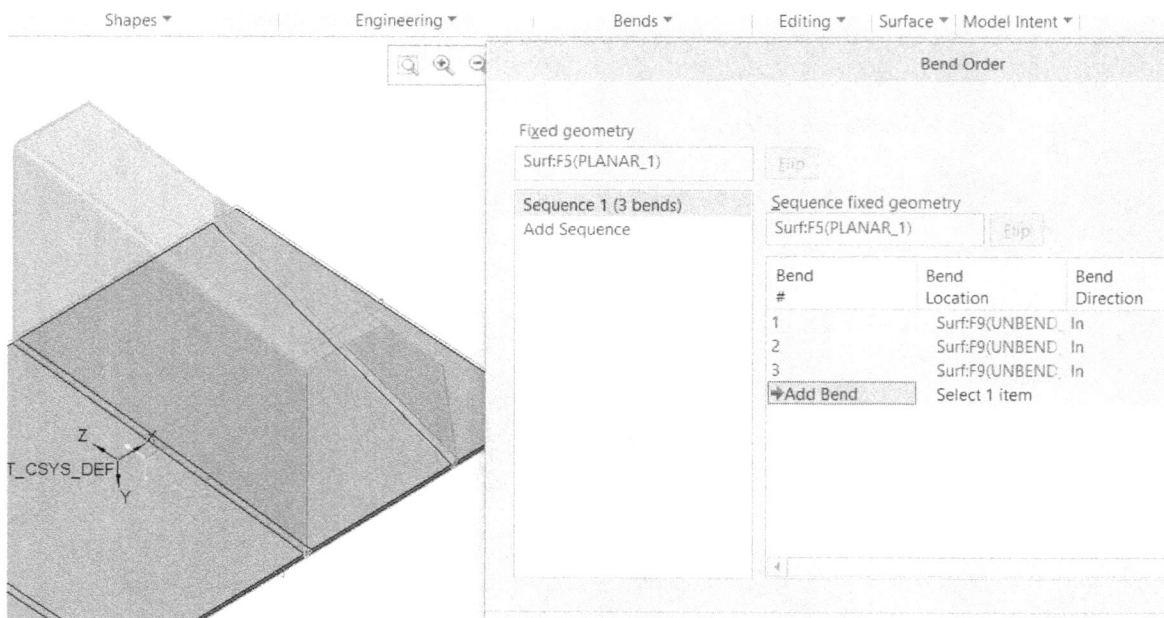

*Figure-69. Selecting bend lines*

• Click on the **OK** button from the dialog box to apply the changes.

## SOLID TO SHEETMETAL

Although there is so much flexibility in working with Sheetmetal environment to create sheetmetal part but still there are some situations where we need to create the part in Part Modeling environment. We can bring the part from Part Modeling to Sheetmetal environment and convert it to sheetmetal part by using the **Convert to Sheet Metal Part** tool in Part Modeling environment and **Conversion** tool in Sheetmetal environment.

The procedure and workflow of converting a solid part into a sheetmetal part is given next.

* Open or create the solid model of part; refer to Figure-70.

*Figure-70. Solid model of part*

* Expand the **Operations** panel in the **Model** tab of Part environment and click on the **Convert to Sheet Metal Part** tool; refer to Figure-71. The **First Wall** contextual tab will be displayed; refer to Figure-72.

*Figure-71. Convert to Sheet Metal Part tool*

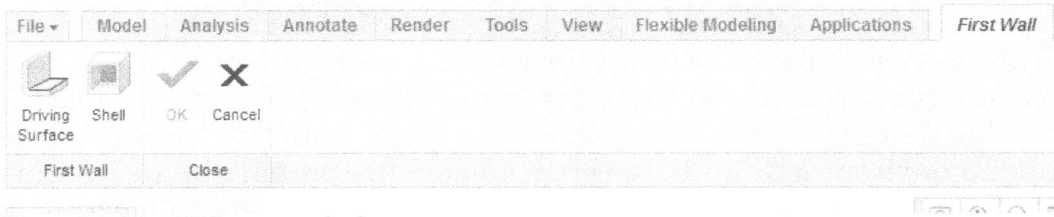

*Figure-72. First Wall contextual tab*

- Click on the **Shell** button from the contextual tab. The **Shell** contextual tab will be displayed with preview; refer to Figure-73.

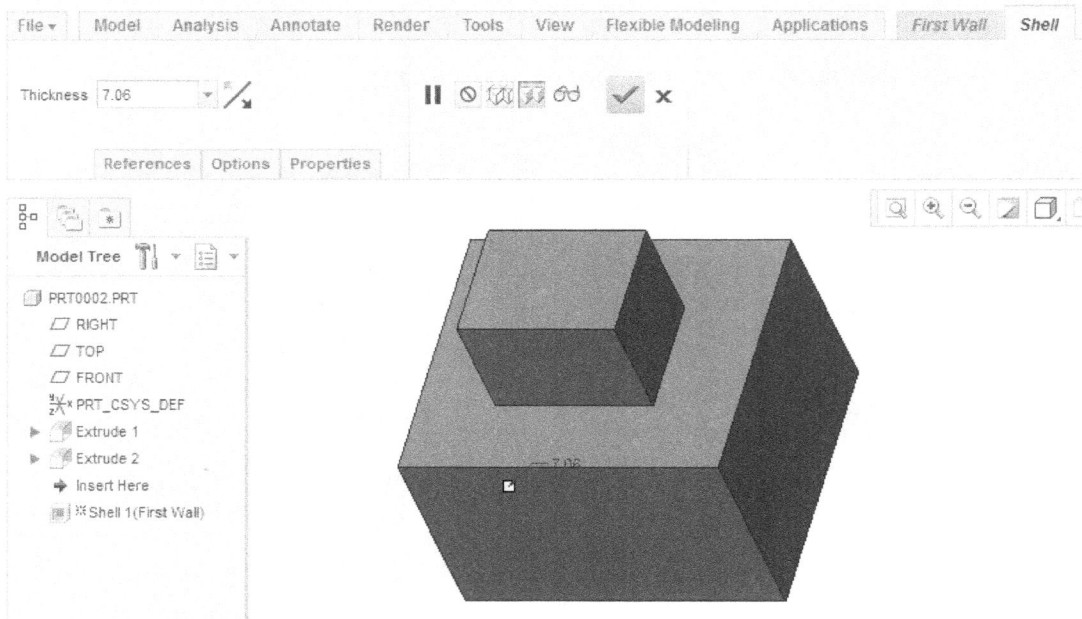

*Figure-73. Shell contextual tab*

- Specify the desired thickness of sheet and click on the **OK** button from the contextual tab. The Sheet Metal environment will be active; refer to Figure-74. (We have done half-way but remember the ultimate goal is Flat Pattern)

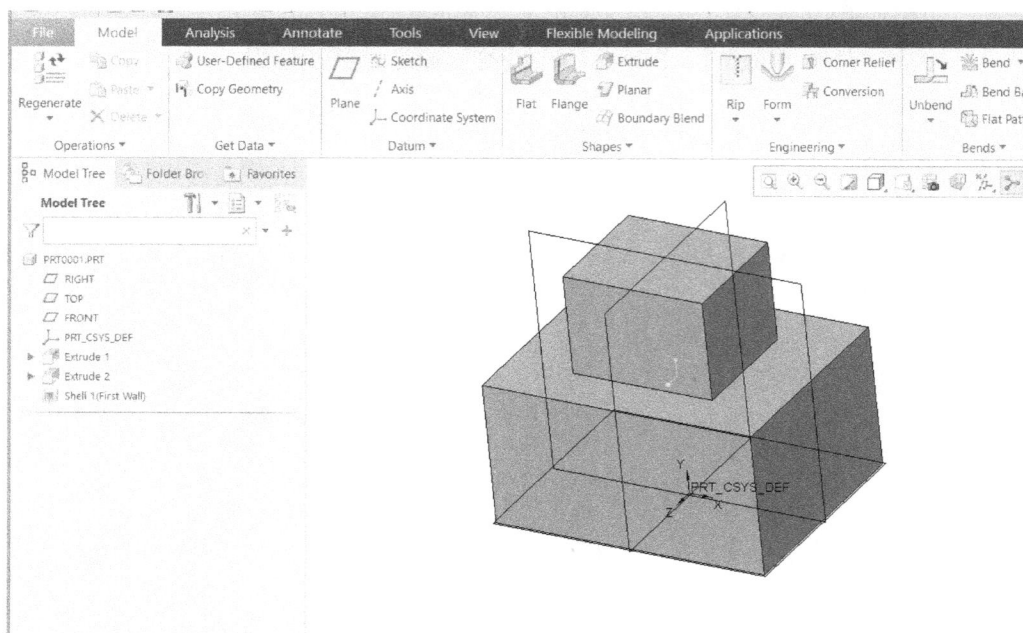

*Figure-74. Part to sheetmetal environment*

- Click on the **Conversion** tool from the **Engineering** panel of the **Ribbon**. The **Conversion** contextual tab will be displayed; refer to Figure-75. (There are four tools in the tab to cut down the edges and faces so that we can unwrap the model for flat pattern.)

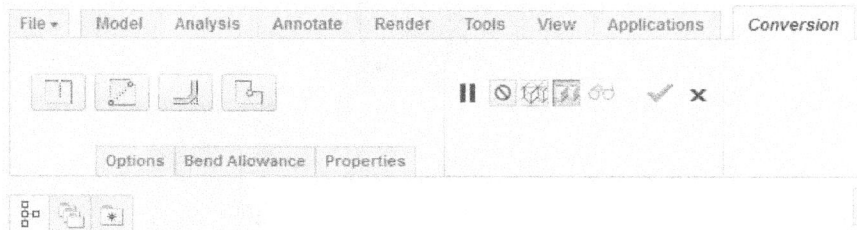

Figure-75. Conversion contextual tab

- Click on the **Options** tab in the contextual tab and make sure the **Inside** option for radius is selected in the drop-down; refer to Figure-76. If **Outside** option is selected then bends as shown in Figure-77 will not be created during conversion.

Figure-76. Inside option for bend radius

Figure-77. Bends that require inside dimensioning

- Click on the **Edge Rip** button ☐ from the contextual tab. The **Edge Rip** contextual tab will be displayed; refer to Figure-78.

Figure-78. Edge Rip contextual tab

- Select the edges as highlighted in Figure-79.

*Figure-79. Edges selected for edge ripping*

- Select the **Gap** option from the **Edge Treatment** drop-down; refer to Figure-80. (Or any other option as you require).

*Figure-80. Edge treatment drop-down*

- Click on the **OK** button from the **Edge Rip** contextual tab. The edges will be ripped allowing the unwrapping of walls; refer to Figure-81. (Now, we got a face stopping the flat pattern creation; refer to Figure-82.)

*Figure-81. Ripped edges*

*Figure-82. Face causing problem*

- Click on the **Rip Connect** button from the **Conversion** contextual tab (displayed again). The **Rip Connect** contextual tab will be displayed with preview of edge rip; refer to Figure-83.

*Figure-83. Rip Connect contextual tab*

- Select two consecutive vertex points of ripped edges as shown in Figure-84. Preview of rip connect will be displayed; refer to Figure-85.

Figure-84. Vertex selected

Figure-85. Preview of rip connect

- Click on the **Placement** tab from the contextual tab and click on the New Set option; refer to Figure-86.
- Select the other two consecutive vertices;refer to Figure-87.

Figure-86. New Set option

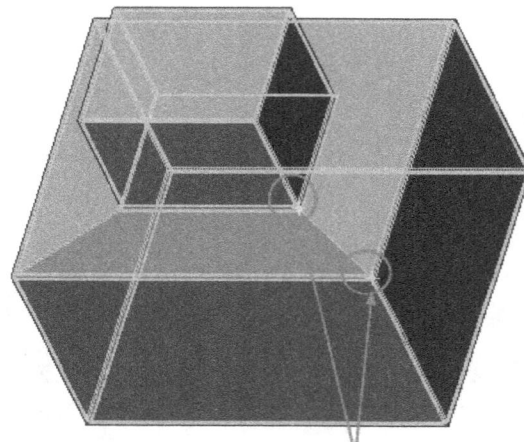

Figure-87. Other vertices selected

- Repeat the procedure for other edge rips. The model will be displayed as shown in Figure-88.

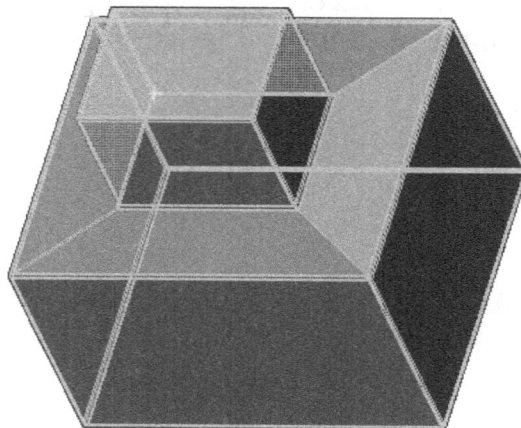

Figure-88. Model after rip connect

- Click on the **OK** button from the **Rip Connect** contextual tab and then from the **Conversion** contextual tab.
- Click on the **Flat Pattern** button from the **Bends** panel in the **Ribbon**. Preview of pattern will be displayed; refer to Figure-89.

*Figure-89. Preview of flat pattern*

Note that you should always check the number of distinct pieces at the bottom bar; refer to Figure-90. If there are more than one distinct pieces then flat pattern will not be created. For example, we have ripped one more edge of the model discussed recently as shown in Figure-91. Now, if you use the **Flat Pattern** tool, an error message will be displayed.

*Figure-90. Distinct pieces*

*Figure-91. One more edge ripped*

# RIP TOOLS

Rip tool is used to make a straight cut in the sheet metal. We use rip when we want to divide a sheet metal piece along an edge or line. There are four tool in **Rip** drop-down of **Engineering** panel in the **Ribbon** to perform this operation; refer to Figure-92. You have used the **Edge Rip** and **Rip Connect** tools in previous topics. Now, we will discuss the use of **Surface Rip** and **Sketched Rip** tools.

*Figure-92. Rip drop-down*

## Surface Rip

The **Surface Rip** tool is used to rip selected surfaces. The procedure to use this tool is given next.

• Click on the **Surface Rip** tool from the **Rip** drop-down in the **Engineering** panel of the **Ribbon**. The **Surface Rip** contextual tab will be displayed; refer to Figure-93. Also, you will be asked to select the surface to be ripped.

*Figure-93. Surface Rip contextual tab*

• Select the surfaces of sheetmetal part that you want to be removed. Preview of the surface rip feature will be displayed; refer to Figure-94.

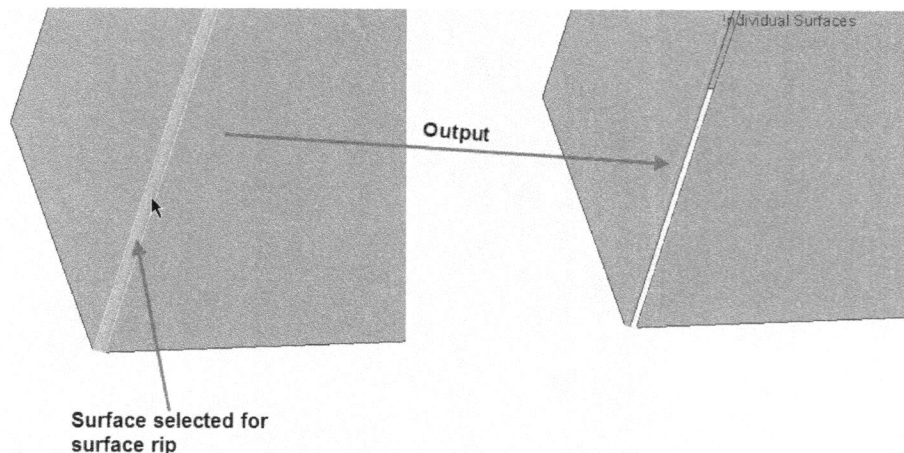

*Figure-94. Surface rip*

• Click on the **OK** button from the **Ribbon** to create the feature.

## Sketched Rip

The **Sketched Rip** tool is used to create rip based on the sketch created. The procedure to use this tool is given next.

- Click on the **Sketched Rip** tool from the **Rip** drop-down in the **Engineering** panel in the **Ribbon**. The **Sketched Rip** contextual tab will be displayed in the **Ribbon**; refer to Figure-95. Also, you will be asked to select a face on which the rip feature is to be created.

*Figure-95. Sketched Rip contextual tab*

- Select the face of sheetmetal part on which you want to create the sketched rip feature. The sketching environment will be displayed.
- Create the sketch for rip feature; refer to Figure-96.

*Figure-96. Sketch for sketched rip feature*

- Click on the **OK** button from the **Sketch** contextual tab. Preview of the sketched rip feature will be displayed; refer to Figure-97.

*Figure-97. Preview of sketched rip feature*

- Set the desired rip side using the flip buttons in the **Ribbon**.

- Click on the **OK** button from the **Ribbon** to create the rip feature.

# CORNER RELIEF

The **Corner Relief** tool is used to apply relief at the corner formed by three joining edges of sheetmetal part. Note that this tool is applicable for placed where corner relief has not been defined earlier like at corner created by two flat walls; refer to Figure-98. The procedure to use this tool is given next.

*Figure-98. Corner created by flat walls*

- Click on the **Corner Relief** tool from the **Engineering** panel in the **Ribbon**. The **Corner Relief** contextual tab will be displayed; refer to Figure-99.

*Figure-99. Corner Relief contextual tab*

- By default, all the corners of sheet metal part are selected. If you want to specify corner relief for a specific corner then click on the manual selection button ⟋ from the **Ribbon**.
- After selecting the corners, select the desired type of corner relief from the drop-down in the **Ribbon**.
- Click on the **Placement** tab and set the parameters for the corner relief in the tab; refer to Figure-100.

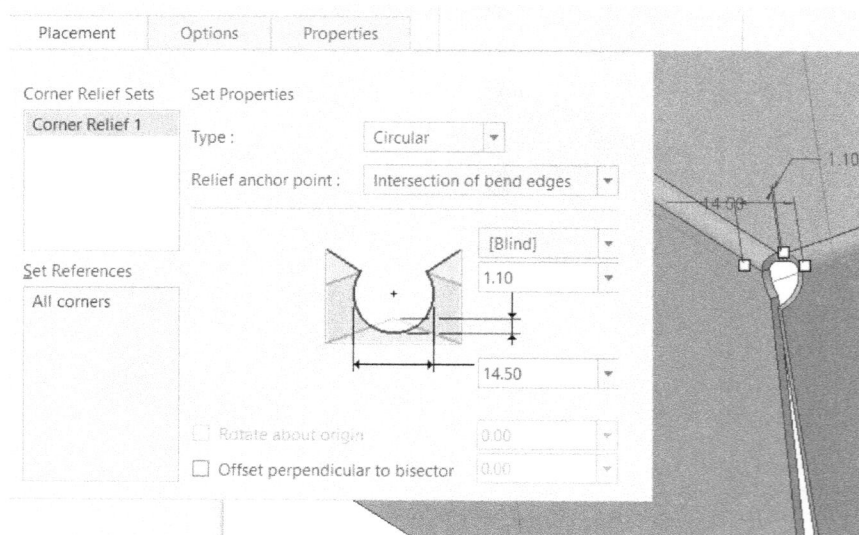

*Figure-100. Specifying parameters for corner relief*

- Click on the **OK** button from the **Ribbon** to create the feature.

# FORMING TOOL

Forming is a process of sheetmetal in which a die/punch is applied to the sheetmetal with mechanical force and impression of die/punch is made on the part. All the sheetmetal forming tools are available in the **Form** drop-down of the **Ribbon**; refer to Figure-101. The procedure to use each forming tool is given next.

*Figure-101. Form drop-down*

## Punch Form Tool

The **Punch Form** tool is used to form punch mark on sheet metal part. The procedure to use this tool is given next.

- Click on the **Punch Form** tool from the **Form** drop-down in the **Engineering** panel of the **Ribbon**. The **Punch Form** contextual tab will be displayed; refer to Figure-102.

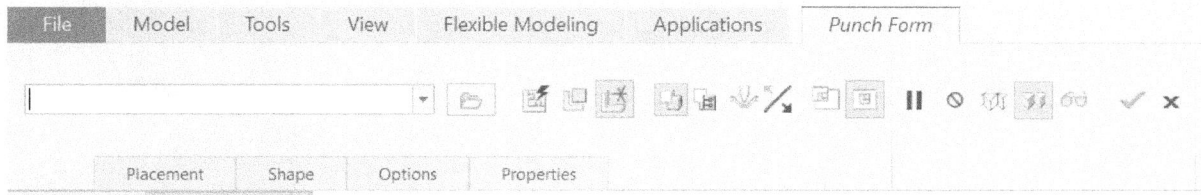

*Figure-102. Punch Form contextual tab*

- Select the desired punch form from the drop-down in the contextual tab or click on the **Open** button and select the part file of punch you want to form on the sheetmetal part. Preview of the punch will be displayed; refer to Figure-103.

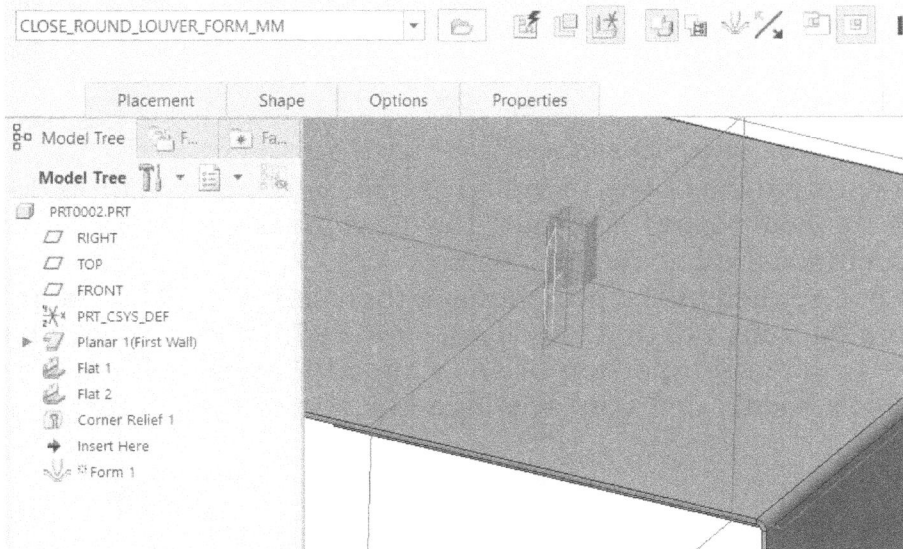

*Figure-103. Preview of punch*

- If you have selected a punch from the drop-down in the **Ribbon** then select the face of sheet metal part on which you want to apply punch. You will be asked to specify distance references. Drag the green handles to edges of face for defining distance; refer to Figure-104.

*Figure-104. Placing punch*

- If you are using a part file as punch using the **Open** button in the **Ribbon** then place the punch at desired location using the assembly constraints; refer to Figure-105.

*Figure-105. Punch placed using assembly constraints*

- If you want to rotate the punch (template punch selected from drop-down in **Ribbon**) then select the **Add rotation about the first axis** check box in the **Placement** tab and specify desired angle of rotation.
- If round is not applied at the punch mark then select the **Placement edges** check box from the **Options** tab and specify desired radius for round (fillet).
- If you want to remove a portion of sheet metal after punching then click in the **Exclude punch model surfaces** selection box and select the respective face; refer to Figure-106.

*Figure-106. Excluding punch model surface*

• Click on the **OK** button from the contextual tab to create the punch mark.

## DIE FORM TOOL

Die creates an opposite impression of punch. In practical, when we apply force on punch then punch presses sheet metal and creates an indent. Opposite to it, in die forming- sheet metal is placed on die and a flat hammer strikes on sheet metal to push sheet metal in die which creates die indent on sheet metal part; refer to Figure-107.

*Figure-107. Die forming and punch forming*

The procedure to use this tool is given next.

• Click on the **Die Form** tool from the **Form** drop-down in the **Engineering** panel of the **Ribbon**. The **Die Form** contextual tab will be displayed. Now, the procedure to use this tool is same as discussed for **Punch Form** tool.

## Sketched Form Tool

The **Sketched Form** tool is used to create a die or punch form mark using sketched shape. The procedure to use this tool is given next.

• Click on the **Sketched Form** tool from the **Form** drop-down in the **Engineering** panel of the **Model** tab in the **Ribbon**. The **Sketched Form** contextual tab will be displayed; refer to Figure-108. Also, you will be asked to select a face to create form tool sketch.

*Figure-108. Sketched Form contextual tab*

- Select the face on which you want to apply the form tool. The Sketching environment will be displayed.
- Create a closed loop sketch to define shape of form tool; refer to Figure-109.

*Figure-109. Sketch created for form tool*

- After creating sketch, click on the **OK** button from the **Sketch** contextual tab. Preview of sketched forming will be displayed; refer to Figure-110.

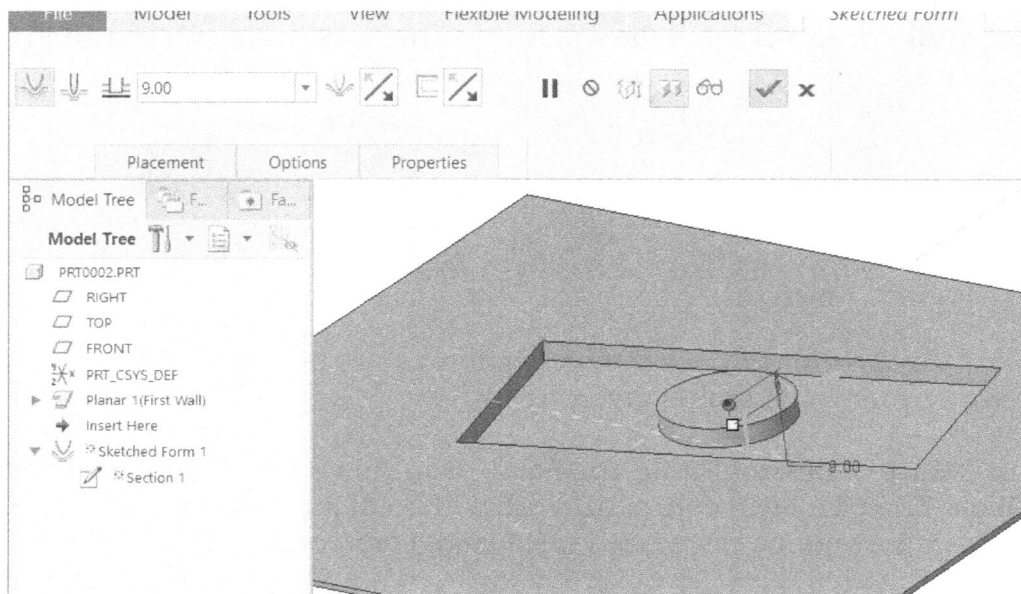

*Figure-110. Preview of sketched forming*

- Click on the **Create a punch** ∨ or **Create a piercing** ⊥ button as required from the **Ribbon**.
- Specify desired depth of form tool in the edit box.
- Set the desired options from the **Options** tab and click on the **OK** button to create the feature.

## Quilt Form Tool

The **Quilt Form** tool is active in the **Form** drop-down when a quilt/surface is available in the modeling area; refer to Figure-111 (You will learn about creating surfaces later in this book).

The procedure to use this tool is given next.

*Figure-111. Surface for forming*

- Click on the **Quilt Form** tool from the **Form** drop-down in the **Engineering** panel of the **Ribbon**. The **Quilt Form** contextual tab will be displayed; refer to Figure-112.

*Figure-112. Quilt Form contextual tab*

- Select the surface/quilt to specify the forming tool and set the direction of forming using the arrows in modeling area; refer to Figure-113.

*Figure-113. Sheetmetal forming using Quilt Form tool*

- Set the desired options in the **Options** tab as discussed earlier and click on the **OK** button.

## Flatten Form Tool

The **Flatten Form** tool is used to flatten the sheet at the locations where forming tools have been applied. The procedure to use this tool is given next.

- Click on the **Flatten Form** tool from the **Form** drop-down in the **Engineering** panel of the **Ribbon**. The **Flatten Form** contextual tab will be displayed and preview of flattening will be displayed; refer to Figure-114.

*Figure-114. Flatten Form contextual tab with preview*

- By default, all the form marks are flattened by this tool but if you want to flatten a specific form mark then click on the **References selected manually** button from the contextual tab in the **Ribbon**. You will be asked to select the form marks to be flattened.
- Select the form marks, preview of flattening will be displayed. If you also want to display projected cuts/holes in flattened sheetmetal then select the **Project cuts and holes** check box from the **Options** tab in the **Flatten Form** contextual tab in the **Ribbon**.
- Click on the **OK** button from the **Ribbon** to create the feature.

## FLAT PATTERN PREVIEW

The **Flat Pattern Preview** tool in **Quick Access Toolbar** is used to check the preview of flat pattern while working on sheet metal part. The procedure to use this tool is given next.

- Click on the **Flat Pattern Preview** tool from the **Quick Access Toolbar** below the **Ribbon**. The **Flat Pattern Preview** window will be displayed with related options; refer to Figure-115.

*Figure-115. Flat Pattern Preview window*

- Select the desired check boxes from **Form geometry** flyout in the toolbar of **Flat Pattern Preview** window to display respective entities in the preview.
- Select the **Bounding Box** button from the toolbar to get dimensions of required blank; refer to Figure-116.

*Figure-116. Blank size displayed*

- You can create the family instance and representation by using the respective buttons in toolbar of **Flat Pattern Preview** window.

Click again on the **Flat Pattern Preview** tool from the **Quick Access Toolbar** to exit the **Flat Pattern Preview** window.

## FLAT PATTERN

The **Flat Pattern** tool is used to create flat pattern of the sheetmetal part which is our ultimate goal for using the software. The procedure to use this tool is same as for **Unbend** tool and options for the tool are also same; refer to Figure-117.

*Figure-117. Flat Pattern contextual tab*

## FLEXIBLE MODELING FOR SHEETMETAL

Flexible modeling is new concept relative to conventional CAD modeling. The tools to perform flexible modeling operations are available in the **Flexible Modeling** tab of the **Ribbon** in Sheetmetal environment; refer to Figure-118. The tools of sheetmetal flexible modeling are discussed next.

*Figure-118. Flexible Modeling tab for sheetmetal*

## Pull Wall Tool

The **Pull Wall** tool is used to stretch a flange or flat wall by specified value. The procedure is given next.

- Click on the **Pull Wall** tool from the **Modify** panel in the **Flexible Modeling** tab of the **Ribbon**. The **Pull Wall** contextual tab will be displayed in the **Ribbon**; refer to Figure-119 and you will be asked to select a wall.

*Figure-119. Pull Wall contextual tab*

- Select the wall to be pulled and specify desired distance value in the edit box of contextual tab. Preview of the feature will be displayed; refer to Figure-120.

*Figure-120. Preview of pull wall feature*

- If you want to include cuts and forms with the selected wall then click on the **Options** tab in the contextual tab and click in the selection box. Now, you can select the cuts and form features that you want to include with the wall to be pulled.
- Using the options in the **Adjacent Conditions** tab, you specify the conditions for features that are adjacent to select wall. Click on the **OK** button to create the feature.

# Edit Bend Tool

The **Edit Bend** tool is used for editing bend dynamically. The procedure to use this tool is given next.

- Click on the **Edit Bend** tool from the **Modify** panel in the **Flexible Modeling** tab of the **Ribbon**. The **Edit Bend** contextual tab will be displayed; refer to Figure-121.

*Figure-121. Edit Bend contextual tab*

- Select the bend that you want to modify. Tools in the contextual tab will become active.

- You can change the angle and bend radius by using the handles displayed in modeling area; refer to Figure-122. Change the other parameters as required using the options in the contextual tab.

*Figure-122. Editing bend*

- Click on the **OK** button to apply modifications.

## Edit Bend Relief Tool

The **Edit Bend Relief** tool is used to edit relief provided for the bend of flanges and other walls. The procedure to use this tool is given next.

- Click on the **Edit Bend Relief** tool from the **Modify** panel in the **Flexible Modeling** tab of the **Ribbon**. The **Edit Bend Relief** contextual tab will be displayed; refer to Figure-123.

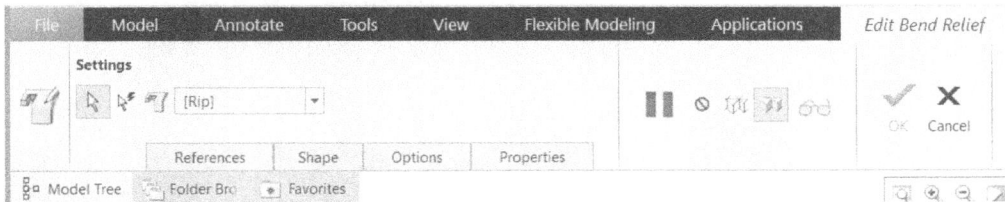

*Figure-123. Edit Bend Relief contextual tab*

- Select the bend whose relief is to be modified and then select the desired option from the drop-down in the contextual tab. Preview of the bend relief will be displayed; refer to Figure-124.

*Figure-124. Preview of editing bend relief*

• Click on the **OK** button to apply modifications.

## Edit Corner Relief Tool

The **Edit Corner Relief** tool is used to edit corner reliefs applied to the sheet metal model. The procedure to use this tool is given next.

• Click on the **Edit Corner Relief** tool from the **Modify** panel in the **Flexible Modeling** tab of the **Ribbon**. The **Edit Corner Relief** contextual tab will be displayed; refer to Figure-125.

*Figure-125. Edit Corner Relief contextual tab*

• Select the corners to be modified. Select the desired corner relief shape from the drop-down in contextual tab and specify the related parameters. Preview of the corner relief will be displayed; refer to Figure-126.

*Figure-126. Preview of editing corner relief*

• Click on the **OK** button to apply the modifications.

In the same way, you can use the **Edit Corner Seam** tool available in the **Modify** panel of the **Flexible Modeling** tab of the **Ribbon**.

## PRACTICE 1

Create the sheet metal model as shown in Figure-127. Dimensions are given in Figure-128.

**Model**

**Flat Pattern**

*Figure-127. Sheet metal model*

Detail A

Thickness of sheet= 1mm
Bend Radius= 5mm

*Figure-128.* *Drawing views*

# PRACTICE 2

Create the sheet metal model as shown in Figure-129. Dimensions are given in Figure-130.

*Figure-129. Practice 2 Model*

Bend Radius 0.100

*Figure-130. Practice 2 Drawing Views*

# SELF ASSESSMENT

1. Which parts are called sheet metal parts?

2. The .......... tool is used to create twisted sheetmetal parts.

3. What is K-factor in case of sheet metal design?

4. While sketching the bend line, you are allowed to create only one line. (T/F)

5. The ........... tool is used to reverse the operation done by **Unbend** tool.

6. Forming is a process of sheetmetal in which a die/punch is applied to the sheetmetal with mechanical force and impression of die/punch is made on the part. (T/F)

7. Write down the difference between punch forming and die forming process.

# FOR STUDENT NOTES

**Answer to Self Assessment Questions:**

1. Those parts which are created by bending, punching, or forming piece of metal sheets 2. Twist 3. The K-Factor in sheet metal work is the ratio of the neutral axis to the material thickness. The K-Factor is used to calculate flat patterns because it is directly related to how much material is stretched during the bend. 4. T 5. Bend Back 6. T 7. Die creates an opposite impression of punch. In practical, when we apply force on punch then punch presses sheetmetal and creates an indent. Opposite to it, in die forming- sheetmetal is placed on die and a flat hammer strikes on sheetmetal to push sheetmetal in die which creates die indent on sheetmetal part

# Chapter 12

# Surface Design

## Topics Covered

The major topics covered in this chapter are:

- *Introduction*
- *Starting with Surface Design*
- *Solid Modeling tools for surfacing*
- *Boundary Blend and Fill surfaces*
- *Trimming, Merging and Offsetting surfaces*
- *Thickening, Solidifying, and Editing surfaces*

# INTRODUCTION

Surface Designing or Surfacing is a technique used to create complex shapes. The most general application of Surfacing can be seen in the object that are made with good aerodynamics like car body, aeroplanes, ships and so on. Sometimes, surfacing is also used to add attractiveness in the models. The models created by surfacing are called surface models and they do not exist in real world. Since, surface models do not have mass properties. So, in the end of surface designing, we are required to convert the surface model into a solid part for manufacturing.

In Creo Parametric, there is no separate environment for creating surface designs. We have all the surfacing tools available in the **Part** Modeling environment; refer to Figure-1. Note that we have also enclosed the tools of **Shapes** panel for surfacing tool. We have used these tools earlier in Solid modeling and now we will use them in surface designing.

*Figure-1. Surfacing tools*

# EXTRUDED SURFACE

The **Extrude** tool can also be used to create the extruded surfaces. Using this tool, we can create flat surface; the procedure is given next.

- Click on the **Extrude** tool from the **Shapes** panel in the **Model** tab of **Ribbon**. The **Extrude** contextual tab will be displayed as discussed earlier in the book.
- Click on the **Extrude as surface** button from the contextual tab.
- Select a plane perpendicular to which you want the surface to be created; refer to Figure-2.
- Create the sketch and click on the **OK** button from the **Sketch** contextual tab. Note that the sketch can be open as well as closed. There is no restriction of closed sketch like in Part modeling.
- On exiting the sketching environment, preview of the extruded surface will be displayed; refer to Figure-3.

Plane selected

TOP:F2(DATUM PLANE)

PRT_CSYS_DEF

*Figure-2. Plane selected*

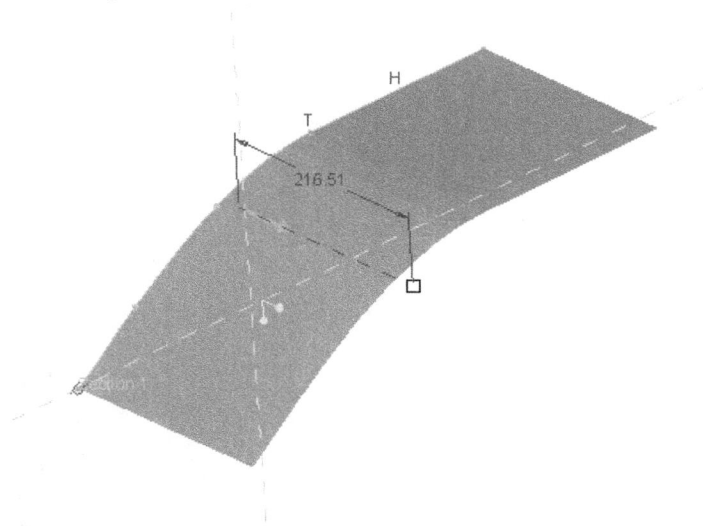

H

T

216.51

*Figure-3. Preview of extruded surface*

- If you have created a closed sketch then you can use the **Capped ends** check box from the **Options** tab in the **Extrude** contextual tab to close the open ends of surface; refer to Figure-4.
- You can also make the surface taper by using the **Add taper** check box as done earlier for solid extrusion.
- Click on the **OK** button from the contextual tab to create the surface.

*Figure-4. Capped ends option*

# OTHER SOLID MODELING TOOLS FOR SURFACING

The other tools in **Shapes** panel can also be used in the same way as discussed in **Chapter 4:3D Modeling Basics** of the book. The only difference while creating the features will be selecting surface modeling button in place of solid modeling button . We can now create surfaces by using **Revolve**, **Sweep**, **Helical Sweep**, **Blend**, **Swept Blend** and **Rotational Blend** tools by following the procedures discussed earlier.

# BOUNDARY BLEND

The **Boundary Blend** tool is one of the most used tools in Creo Parametric for surfacing. Using this tool, you can create surface with the help of sketched curves or edges of other entities. The procedure to use this tool is given next.

- Click on the **Boundary Blend** tool from the **Surfaces** panel in the **Model** tab of **Ribbon**. The **Boundary Blend** contextual tab will be displayed; refer to Figure-5.

*Figure-5. Boundary Blend contextual tab*

- There are two curve collectors in the contextual tab; refer to Figure-5. Click in the first collector and select the curves in first direction. Preview of surface will be displayed; refer to Figure-6.

*Figure-6. Preview of boundary blend with 1 direction curves*

- To add curves of 2nd direction, click in the **2nd Direction** collector and select the curves. Preview of boundary blend will be displayed; refer to Figure-7.

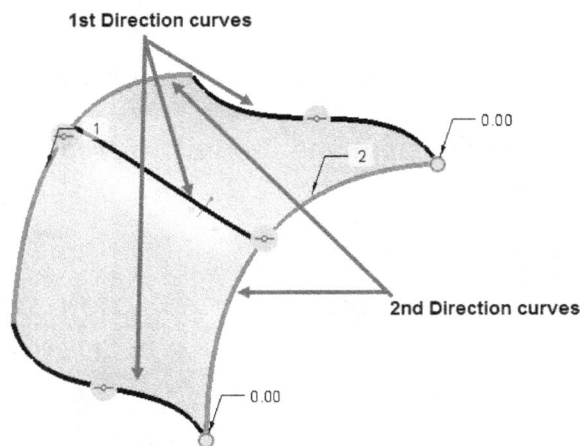

*Figure-7. Boundary blend with curves in both direction*

- You can also use an influencing curve to change the shape of boundary blend. To do so, click on the **Options** tab in the **Boundary Blend** contextual tab. The options will be displayed as shown in Figure-8.

*Figure-8. Options tab*

- Click in the **Influencing curves** collector and select the sketched curve that you want to use for influencing the surface. Preview will be displayed; refer to Figure-9.

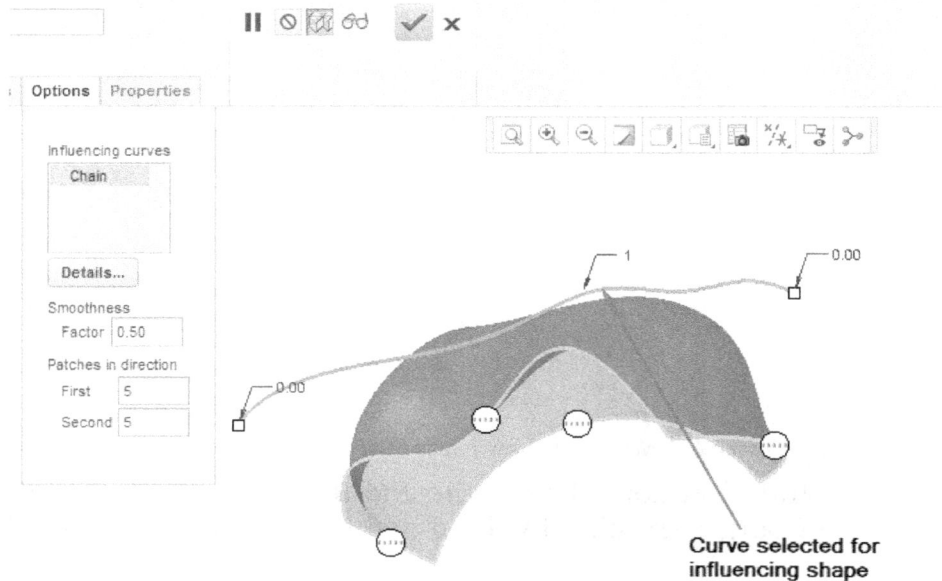

*Figure-9. Boundary blend with influencing curve selected*

- Click on the **OK** button from the contextual tab to create the surface.

Note that you can also use the edges of solid models to create boundary blends; refer to Figure-10.

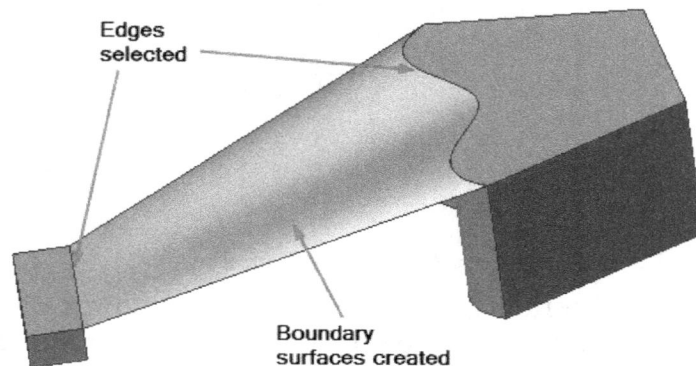

*Figure-10. Boundary blend using edges of solids*

# FILL

The **Fill** tool is used to fill the void areas of solids and surfaces. It works with the help of closed sketches. Most of the time, we use it to create planar surfaces with irregular shapes but we can also use it along with **Project** tool to fill gaps between other solids/surfaces. The procedure for both the conditions is given next.

## Creating Planar Surfaces with Fill

- Click on the **Fill** tool from the **Surfaces** panel in the **Model** tab of the **Ribbon**. The **Fill** contextual tab will be displayed; refer to Figure-11.

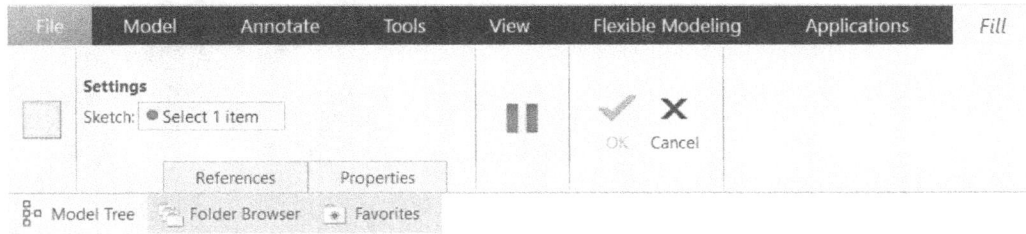

*Figure-11. Fill contextual tab*

- Click on the **References** tab and then click on the **Define** button from it.
- Select the desired plane and create the sketch for planar surface. Note that the sketch must be closed one loop.
- After creating the sketch, click on the **OK** button from the **Sketch** contextual tab. The preview of the surface will be displayed; refer to Figure-12.

*Figure-12. Preview of fill surface*

- If you want to fill the gap on a planar surface, then project the edges of the gap using the **Project** tool from the **Sketching** panel to form a closed loop; refer to Figure-13.

*Figure-13. Edges projected to fill gap*

- Click on the **OK** button from the **Fill** contextual tab to create the surface.

## TRIMMING SURFACES

Trimming is an important step while creating surface models. As we trim sketched entities while sketching, in the same way we can trim the surfaces intersecting with other surfaces. The procedure for trimming surfaces is given next.

- Select the surface that you want to trim, the **Trim** tool will become active in the **Editing** panel of the **Ribbon**; refer to Figure-14.

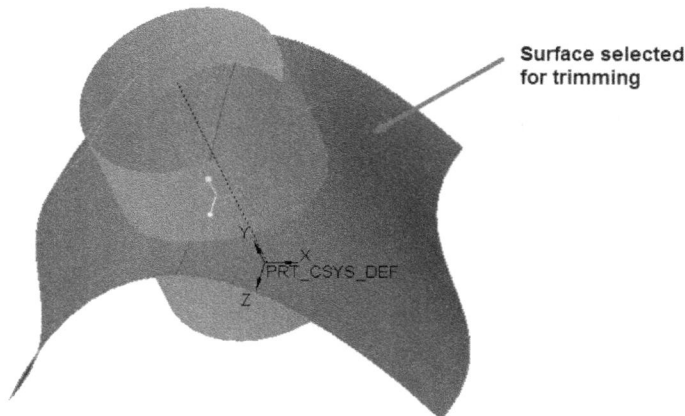

*Figure-14. Surface selected for trimming*

- Click on the **Trim** tool. The **Surface Trim** contextual tab will be displayed; refer to Figure-15.

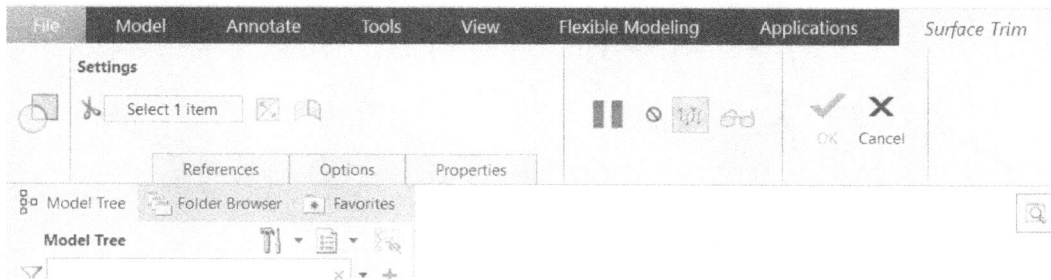

*Figure-15. Surface Trim contextual tab*

- Click on the **References** tab and select the other surface which you want to use as trimming tool; refer to Figure-16.

*Figure-16. Selecting trimming surface*

- Click on the **Flip** button to change the side of surface which you want to remain. Note that the portion of surface which is displayed with orange dotted net will remain and other portion will be trimmed. Clear the **Keep trimming surface** check box from the **Options** tab in **Surface Trim** contextual tab if you want to remove the trimming tool.
- Click on the **OK** button from the contextual tab. The other side of surface will be trimmed; refer to Figure-17.

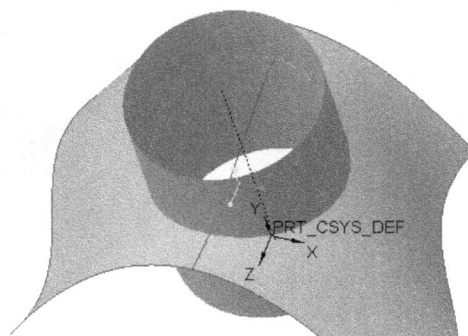

*Figure-17. Surface after trimming*

Note that you can use projected sketches as trimming tool to cut a surface as shown in Figure-18.

Output →

**Surface selected**
**for trimming**

**Projected curve selected**
**as trimming tool**

*Figure-18. Surface trimmed by curve*

# MERGING SURFACES

There can be various purposes of merging two or more surfaces; like applying same thickness to all the surfaces later, creating a closed surface to form solid, and so on. We will learn to form solid using surfaces later in this chapter. The procedure to merge surfaces is given next.

- Click on the **Merge** tool once you have selected two intersecting surfaces. The **Merge** contextual tab will be displayed along with the preview of merged surfaces; refer to Figure-19.
- Using the two **Flip** buttons ⅄ in the contextual tab, you can change the sides of surfaces. The direction in which the Pink arrows point will remain and others will be trimmed.
- Click on the **OK** button from the **Merge** contextual tab to merge the surfaces.

*Figure-19. Merge contextual tab with preview of merged surfaces*

# OFFSETTING SURFACES

The **Offset** tool in surfacing works in the same way as it does in sketching for sketched entities. The **Offset** tool is used to create copy of the selected entity at a specified distance. The procedure to offset surfaces is given next.

* Select the surface that you want to offset and click on the **Offset** tool from the **Editing** panel in the **Ribbon**. Preview of the offsetted surface will be displayed along with the **Offset** contextual tab in **Ribbon**; refer to Figure-20.

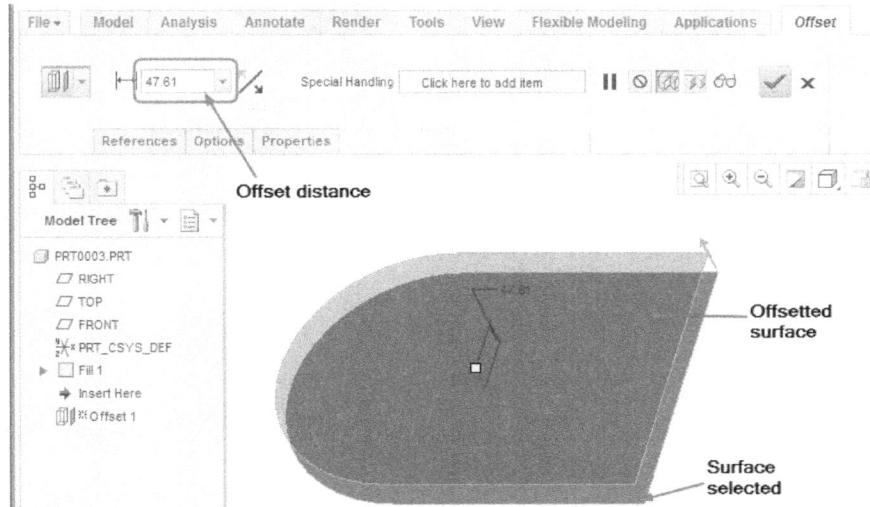

*Figure-20. Preview of Offset*

* The offset preview display in above figure is the most common use of **Offset** tool. But, there are three more options in which we can apply the **Offset** tool in surfacing; **Offset With Draft**, **Offset With Expand**, and **Offset With Replace Surface**. These options are available in the **Feature type** drop-down in the contextual tab; refer to Figure-21.

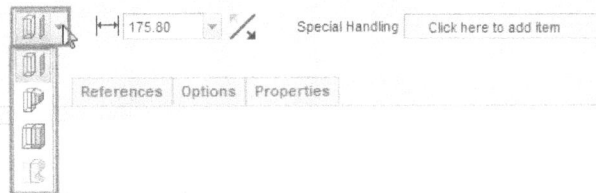

*Figure-21. Feature type drop-down*

## With Draft Feature

* Click on the **With Draft Feature** option from the **Feature type** drop-down. You will be asked to select or create a sketch.
* Select the closed sketch if you have in modeling area otherwise click on the **References** tab and select the **Define** button. You will be asked to select the sketching plane.
* Select the plane and create a closed sketch in such a way that it intersects with the surface; refer to Figure-22.

Figure-22. Sketch created

- After creating sketch, click on the **OK** button from the **Sketch** contextual tab. The preview will be displayed; refer to Figure-23.

Figure-23. Preview of Offset with draft feature

- Specify the desired height and draft angle in the edit boxes available in the contextual tab and click on the **OK** button from the contextual tab. The surface will be created with offsetted region coming out from the base surface; refer to Figure-24.

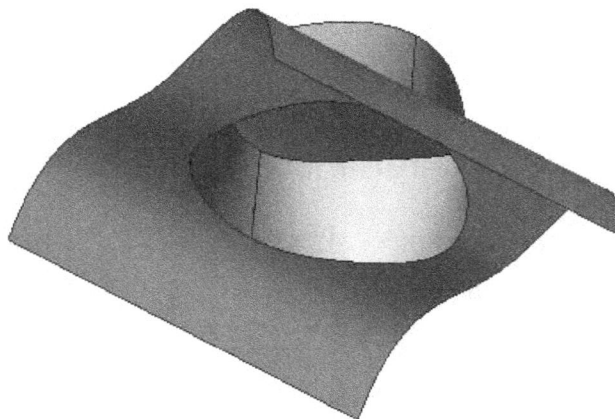

Figure-24. Offsetted surface viewed from bottom

## Expand Feature

You can offset the surface in same way as done using **With Draft Feature** option but without applying draft, by using the **Expand Feature** option. The procedure is same as for **With Draft Feature** option.

## Replace Surface Feature

The **Replace Surface Feature** option is used to replace the face of a solid model with an intersecting surface. In other words, the selected face takes the shape of surface selected as replacement. The procedure to use this option is given next.

- Select the face of a solid model; refer to Figure-25 and then click on the **Offset** tool from the **Editing** panel of the **Ribbon**. The **Offset** contextual tab will be displayed.

*Figure-25. Face of solid model selected for replacement*

- Select the **Replace Surface Feature** option from the **Feature type** drop-down in the contextual tab. You are asked to select the replacement surface.
- Select the replacement surface which is intersecting with the model parallel to replacement face. Preview of replaced surface will be displayed; refer to Figure-26.

*Figure-26. Preview of replaced surface*

- Click on the **OK** button from the contextual tab to create the replacement.

# THICKEN

The **Thicken** tool is used to add desired thickness to a surface. This tool used once you have performed all the surfacing operations. The procedure to use this tool is given next.

- Select the surface to which you want to apply thickness and then click on the **Thicken** tool from the **Editing** panel in the **Ribbon**. Preview of thickened surface will be displayed along with the **Thicken** contextual tab; refer to Figure-27.

*Figure-27. Preview of Thickened surface*

- Click in the edit box in **Ribbon** and specify the desired thickness.
- Click on the **OK** button to create the feature. Note that you can also remove material by using the **Remove Material** button from the contextual tab while applying thickness to surface if a solid model already exists which is intersecting which surface.

# SOLIDIFY

The **Solidify** tool is used to covert closed surfaces into a solid. In other words, this tool fills material in the volume closed by surfaces. Note that the surfaces must be merged by using the **Merge** tool. The procedure to use this tool is given next.

- Select the merged surfaces forming closed volume; refer to Figure-28 and click on the **Solidify** tool from the **Editing** panel of the **Ribbon**. The **Solidify** contextual tab will be displayed.

Figure-28. Merged surfaces selected

- Click on the **OK** button from the contextual tab. Solid will be created using the surface boundaries.

## EDITING SURFACES USING CONTROL POINTS

We can modify the shape of a surface/face by using the control points. **This operation is generally done after creating solid from surface.** The tools to do so are available in the **Style** contextual tab. The procedure to modify surface/face using control points is given next.

- Click on the **Style** tool from the **Surfaces** tab in the **Model** tab of **Ribbon**. The **Style** contextual tab will be displayed; refer to Figure-29.

Figure-29. Style contextual tab

- Click on the **Surface Edit** tool from the **Surface** panel in the contextual tab. The **Style: Surface Edit** contextual tab will be displayed; refer to Figure-30.

Figure-30. Style: Surface Edit contextual tab

- Select the surface/face of model that you want to modify. Control points will be displayed on the surface/face; refer to Figure-31.

*Figure-31. Face with control points*

- Increase the number of control points by using the **Max Rows** and **Columns** spinners in the contextual tab.
- Drag the control points from the modeling area to modify the shape of surface; refer to Figure-32.

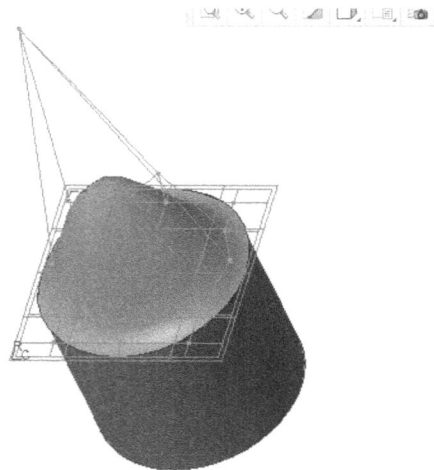

*Figure-32. Surface editted*

- Click on the **OK** button from the contextual tab to create a style surface with desired shape.
- Click on the **OK** button from the **Style** contextual tab to create the feature. You will notice that it has become a separate surface; refer to Figure-33. If you want this surface to be applied on solid model, use the **Replace Surface Feature** option in the **Offset** tool. The model after replacing the face with style surface will be displayed as shown in Figure-34.

*Figure-33. Style surface created*

*Figure-34. Model after replacing the face*

## VERTEX ROUND

The **Vertex Round** tool is used to apply round at the sharp vertex of a surface. The procedure to use this tool is given next.

- Click on the **Vertex Round** tool from the expanded **Surfaces** panel of the **Model** tab in the **Ribbon**. The **Vertex Round** contextual tab will be displayed and you will be asked to select vertices to be rounded.
- Select the vertices. Preview of the round will be displayed; refer to Figure-35.

*Figure-35. Preview of vertex round*

- Specify the desired value of radius and click on the **OK** button from the **Ribbon**.

# PRACTICAL 1

Create the model of helmet glass as shown in Figure-36. The dimensions of the model are given in Figure-37.

*Figure-36. Practical1 model*

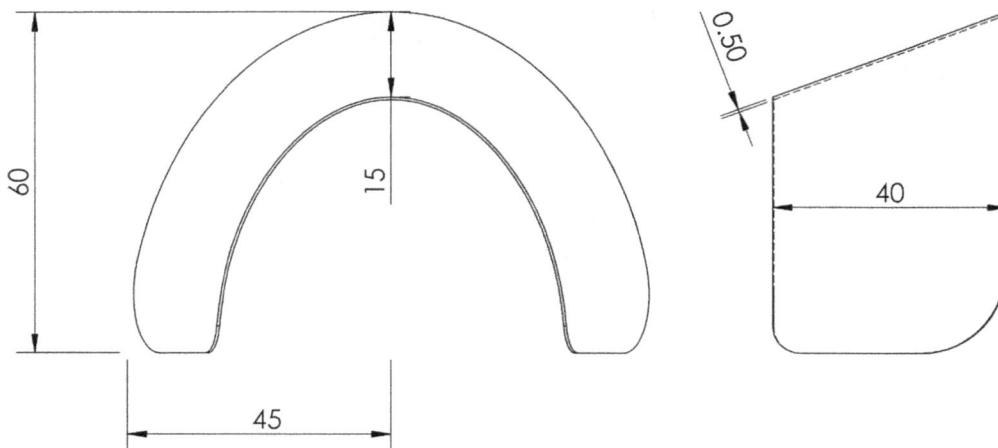

*Figure-37. Practical1 drawing*

The model displayed is having very low thickness and its having complex 3D shape. So, its a good idea to use surfacing in this case.

We can create this model by Blend tool easily. For that we need to have two sketches. Start the Part Modeling environment of Creo Parametric.

## Creating first sketch

- Click on the **Sketch** tool from the **Datum** panel in the **Model** tab of the **Ribbon**. The Sketch dialog box will be displayed and you will be prompted to select a sketching plane.
- Select the **Top** plane and click on the **Sketch** button. The sketching environment will display.
- Click on the **Center and Axis Ellipse** tool from the **Ellipse** drop-down in the **Sketching** panel of the **Ribbon** and draw an ellipse as shown in Figure-38.

*Figure-38. Ellipse to be drawn*

- Trim the ellipse by using the **Delete Segment** tool in such a way that upper half of ellipse is left; refer to Figure-39.
- Click on the **OK** button from **Sketch** contextual tab to exit the sketch.

*Figure-39. Trimmed ellipse*

## Creating blended surface

- Click on the **Blend** tool from expanded **Shapes** panel in **Model** tab of the **Ribbon**. The **Blend** contextual tab will be displayed; refer to Figure-40.

*Figure-40. Blend contextual tab*

- Select the **Blend as surface** button from the contextual tab.
- Click on the **Sections** tab and select the sketch created. You are asked to specify offset distance for second sketch.
- Specify the offset distance as **40** and click on the **Sketch** button; refer to Figure-41.

*Figure-41. Offset distance specified*

- Click on the **Offset** tool from the **Sketching** panel in the **Ribbon** and selected the earlier created sketch.
- Specify the value of offset as **-15** if offset direction is **outward** and **15** if offset direction is **inward** in the input box displayed; refer to Figure-42.

*Figure-42. Sketch offsetted*

- Click on the **OK** button from the Input box and **Close** button from the **Type** Menu Manager displayed to create the offset curve; refer to Figure-43. Note that the starting points of curves should be properly aligned. After aligning the starting points the sketch should display like Figure-44.

*Figure-43. Offset sketch created*

*Figure-44. Sketch after aligning starting points*

- Exit the sketching environment by pressing **OK** button. Preview of blended surface will be displayed; refer to Figure-45.

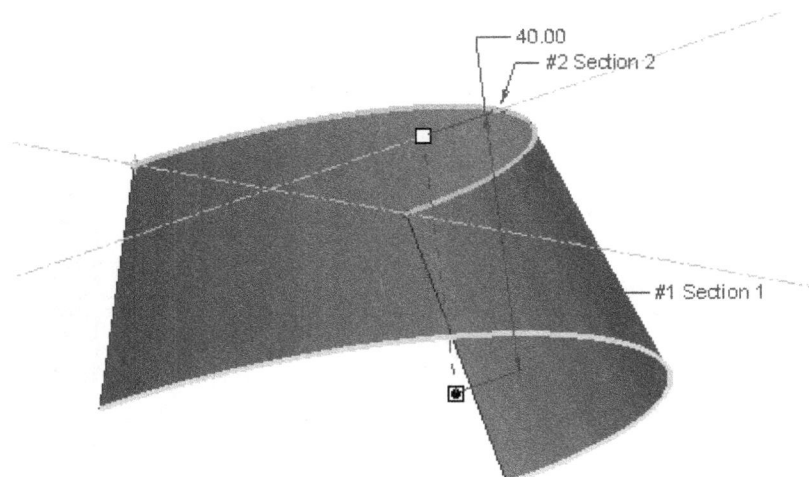

*Figure-45. Blended surface*

- Click on the **OK** button from the **Blend** contextual tab to create the surface.

## Thickening surface and applying fillet

- Click on the **Thicken** tool from the **Ribbon** and select the surface. Preview of thickened surface will display.
- Enter the thickness value as **0.5**.
- Click **OK** to create the solid.
- Click on the **Fillet** tool and apply suitable fillets at the small edges on the corners.

# PRACTICE 1

Create the surface model of tank as shown in Figure-46. The dimensions of the model are given in Figure-47.

Figure–46. Practice1 model

Figure–47. Practice1 drawing

# PRACTICE 2

Create the surface model of car bumper as shown in Figure-48. The dimensions of the model are given in Figure-49. **Assume the missing dimensions. (Hint:** You will need the **Style** tools to modify the surface after using **Swept Blend** tool.**)**

Figure-48. Practice 2 model

Figure-49. car bumper

# SELF ASSESSMENT

1. The tools to create surface are available in :

(a) Part Modeling Environment     (b) Drawing Environment
(c) Assembly Environment          (d) Manufacturing Environment

2. The .......... tool is used to create surface using sketched curves or edges of other entities as boundaries.

3. The .................... option is used to replace the face of a solid model with an intersecting surface.

4. The ............. tool is used to covert closed surfaces into a solid.

5. The .............. tool is used to apply round at the sharp corner of a surface.

**Answers to Self Assessment Questions:**
1. (a) 2. **Boundary Blend** 3. **Replace Surface Feature** 4. **Solidify** 5. **Vertex Round**

# Chapter 13

# Drawing

## Topics Covered

The major topics covered in this chapter are:

- **Introduction**
- **Starting Drawing**
- **Inserting new sheets**
- **Placing base view**
- **Creating Projection view, Detail view, Section View, Broken Views and so on**
- **Applying dimensions to views**
- **Creating Geometric tolerance, surface finish, and datum symbols**
- **Creating Tables, Bill of Materials, and Balloons**

## INTRODUCTION

We create models and assemblies in Creo Parametric for manufacturing. These models are not useful for manufacturer until they give complete **Product Manufacturing Information** (PMI) to them. The manufacturer will not be able to find the dimension of product by barely looking at the model that does not display any dimension or geometric tolerances in a computer system/gadget. So, we need to represent the models and assemblies in such a way that all the dimensions and geometric tolerances are available for easy interpretation. In latest CAD software, you can use **Model Based Definition** tools to put Product Manufacturing Information directly on model or assembly and then you can use it at shop floor for manufacturing (You will learn about MBD later). But, still most of the industries use drawings printed on papers for their manufacturing. To represent real objects on paper, we use engineering drawings at the workshops. In this chapter, we will be discuss about the tools available in Creo Parametric related to drawing creation.

## STARTING DRAWING

- Start Creo Parametric and click on the **New** button from the **Quick Access Toolbar**. The **New** dialog box will be displayed; refer to Figure-1.

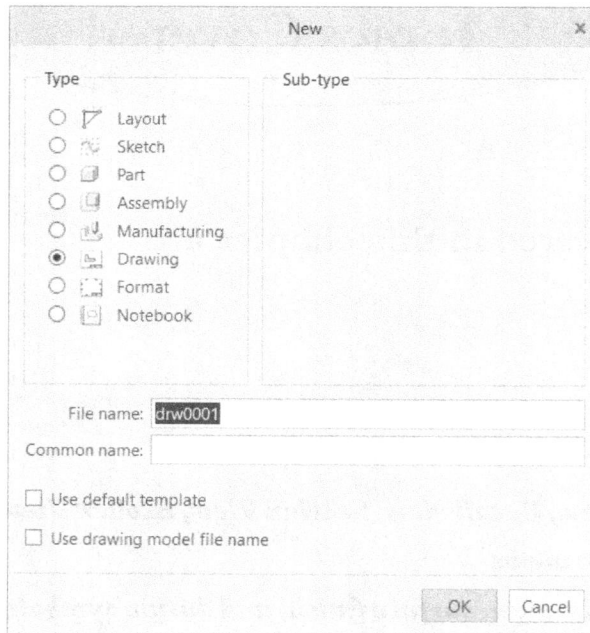

*Figure-1. New dialog box*

- Select the **Drawing** radio button from the **Type** area of the dialog box. Clear the **Use default template** check box to manually select the template. Select the **Use drawing model file name** check box to use the name of model imported in drawing as drawing file name and click on the **OK** button from the dialog box. The **New Drawing** dialog box will be displayed; refer to Figure-2.

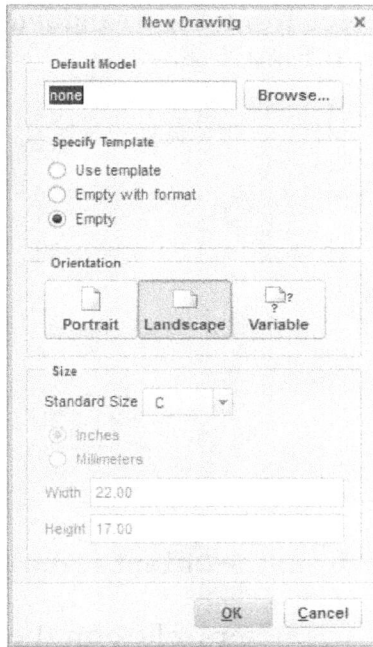

*Figure-2. New Drawing dialog box*

- Click on the **Browse** button from the **Default Model** area of the dialog box. The **Open** dialog box will be displayed; refer to Figure-3.

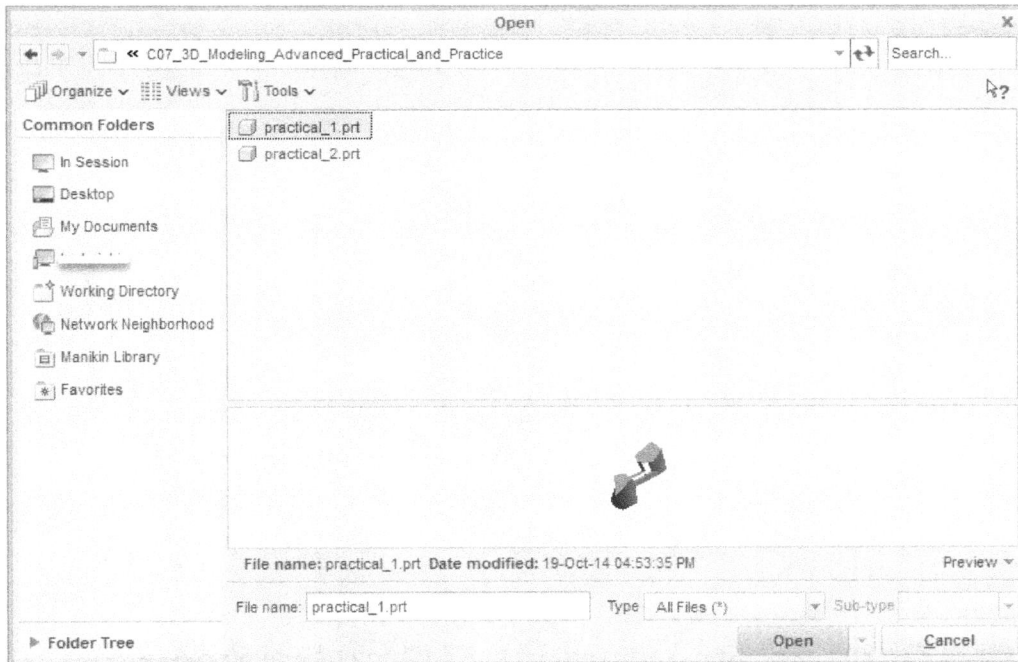

*Figure-3. Open dialog box*

- Select the part/assembly model for which you want to create the drawing and click on the **Open** button from the dialog box.
- Select the desired orientation of page from the **Orientation** area of the dialog box; refer to Figure-4.

*Figure-4. Orientation area of dialog box*

- Select the desired size of sheet from the **Standard Size** drop-down in the **Size** area of the dialog box; refer to Figure-5.

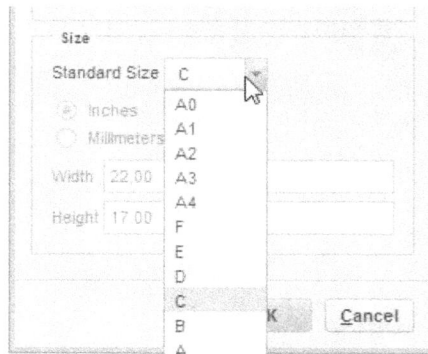

*Figure-5. Sheet size drop-down*

- Click on the **Variable** button from the **Orientation** area of the dialog box to specify custom size of the sheet.
- After specifying the desired parameters, click on the **OK** button from the dialog box. The Drawing environment of Creo Parametric will be displayed; refer to Figure-6.

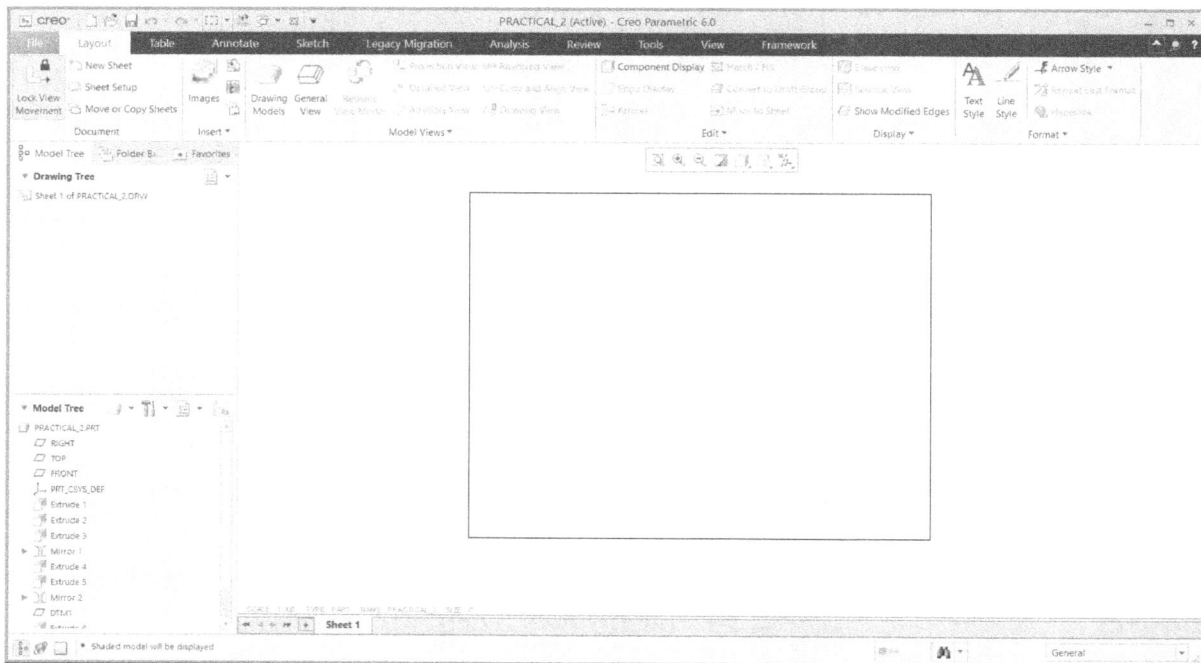

*Figure-6. Drawing environment*

## Drawing environment of Creo Parametric

There is a lot we do using Drawing environment of Creo Parametric so that a person on manufacturing facilities can manufacture the part without any doubt. Following are the general things that we do using the Drawing environment:

- We insert various views of model in the drawing sheet.
- We dimension the views both geometrically and dimensionally.
- We create Tables in the drawing.
- We create Bill of Materials and Balloons.
- We insert images and other data for presentation purpose.

We will start with inserting various type of views in the drawing sheet. But, before that we will learn to create more drawing sheets.

## INSERTING DRAWING SHEETS

Drawing sheet is the base for various drawing views. Without drawing sheet, we cannot create drawings in Creo. The drawing sheets are displayed in **Drawing Sheet** tabs at the bottom of the drawing area; refer to Figure-7. Here, Sheet 1 is the name of the sheet. Double click on it to change the name. The procedure to insert new drawing sheets is given next.

*Figure-7. Drawing Sheet tab*

*   Click on the **Layout** tab of the **Ribbon** if not selected by default. The options related to sheet will be displayed in the **Document** panel of the **Ribbon**; refer to Figure-8.

*Figure-8. Document panel in Layout tab*

*   Click on the **New Sheet** tool from the **Document** panel. A new sheet will be added in the **Sheet** tabs at the bottom with the name **Sheet 2**.
*   By default, the new sheet has the same parameters as we have set while starting the drawing environment. To change the parameters of the sheet, click on the **Sheet Setup** button from the **Document** panel of the **Layout** tab in **Ribbon**. The **Sheet Setup** dialog box will be displayed; refer to Figure-9.

*Figure-9. Sheet Setup dialog box*

*   Click in the cell under **Format** column of the dialog box. It will change to a drop-down. Select the desired size for the sheet; refer to Figure-10. You can select the **Custom Size** option and specify the sheet size as required in the edit boxes of **Size** area in the dialog box.

*Figure-10. Format drop-down*

- Click on the **OK** button from the dialog box to apply the changes.

## ADDING MODEL TO THE DRAWING

It might be possible that you have selected a wrong model and have started the drawing or you have not selected any model but are in the drawing environment. So, the steps to add or replace the model for drawing are given next.

- Click on the **Drawing Models** tool from the **Model Views** panel of the **Ribbon**. The **DWG MODELS** Menu Manager will be displayed; refer to Figure-11.

*Figure-11. DWG MODELS Menu Manager*

- Click on the **Add Model** option if you have not selected any model while starting the drawing environment. The **Open** dialog box will be displayed as discussed earlier.
- Select the part/assembly file for which you want to create the drawing views and click on the **OK** button from the dialog box. The model will be fetched in the drawing memory.
- If you want to change the model, then click on the **Del Model** option from the **Menu Manager**. A confirmation message box will be displayed; refer to Figure-12.

*Figure-12. Confirmation box*

- Click on the **Yes** button to accept the deletion.

- Click on the **Drawing Models** tool again from the **Model Views** panel and then click on the **Add Model** option. Rest of the procedure has already been discussed.
- After adding the model, click on the **Done/Return** option from the Menu Manager.

## PLACING THE FIRST VIEW (BASE VIEW)

First View or **Base View** is an independent view and all other views are placed with reference to this view either directly or indirectly. The Base view is placed by using the **General View** tool. The procedure to place base view is given next.

- Click on the **General View** tool from the **Model Views** panel of the **Ribbon**. The **Select Combined State** dialog box will be displayed; refer to Figure-13.

*Figure-13. Select Combined State dialog box*

- Select the **No Combine State** option from the dialog box and select the **Do not prompt for combined state** check box. Using the **No Combine State** option, you specify that no user defined states are required in the drawing mode.
- Click on the **OK** button from the dialog box. You are asked to specify the center location for the base view.
- Click in the drawing area at desired location to place the base view. As soon as you click in the drawing area, the default view of model and **Drawing View** dialog box are displayed; refer to Figure-14.

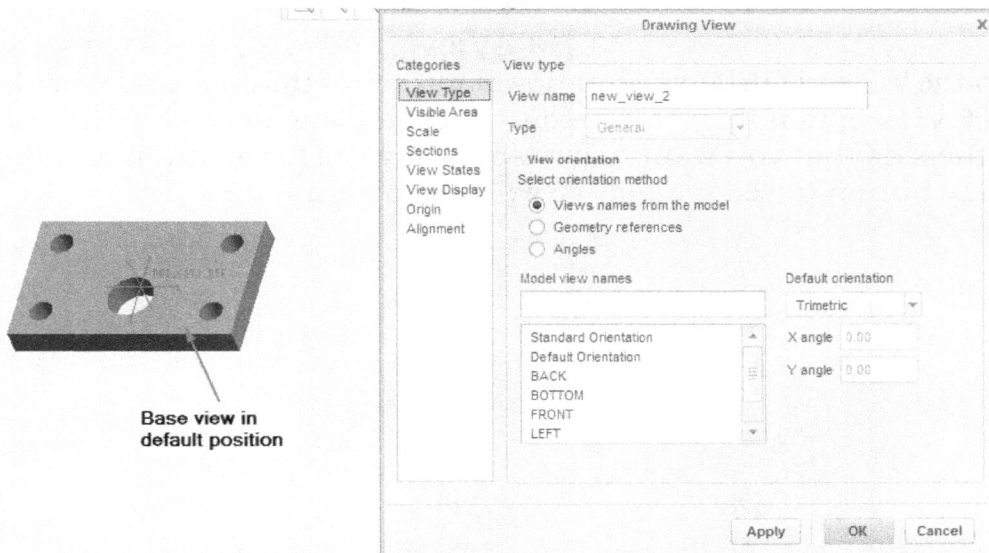

*Figure-14. Base view with Drawing View dialog box*

# View Type options

- Click in the **View name** edit box in the dialog box and specify the desired name for the view, like Front view or Top view.
- Select the desired model view from the **Model view names** list box. Like, FRONT or TOP. You can select the **Default Orientation** option and specify it as **Trimetric**, **Isometric** or user defined by using the option in the **Default orientation** drop-down in the dialog box.
- After selecting the desired view, click on the **Apply** button from the dialog box.

# Visible Area options

- To clip the view or, display half or partial view; click on the **Visible Area** option from the **Categories** list box in the dialog box. The dialog box will be displayed as shown in Figure-15.

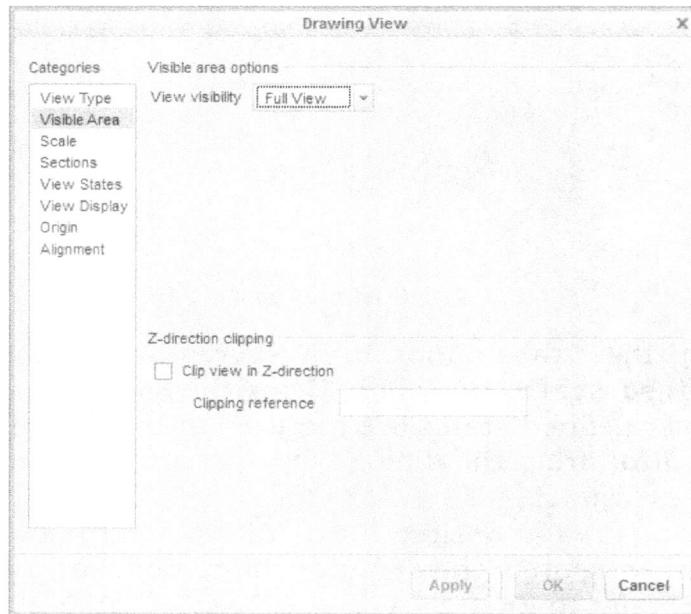

*Figure-15. Drawing View dialog box with Visible Area options*

# Half-View

- Click in the **View visibility** drop-down and select the desired option. If you select the **Half view** option then you need to select a plane normal to the selected view. Select the side that you want to keep by using the **Flip** button and then click on the **Apply** button to see the half-view; refer to Figure-16.

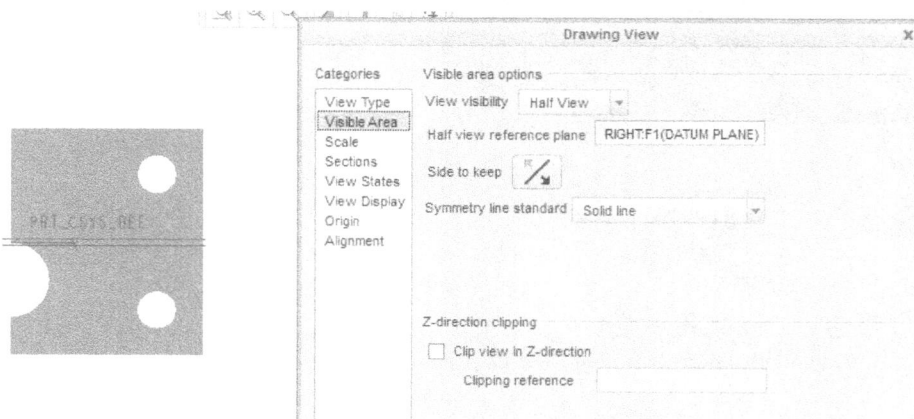

*Figure-16. Half-view of model*

## Partial View

- To create the partial view, click on the **Partial View** option from the **View visibility** drop-down. You are asked to select a reference point. Select a point and then create a spline around it; refer to Figure-17. Click on the **Apply** button from the dialog box to check the partial view; refer to Figure-18.

Figure-17. Spline created for partial view

Figure-18. Partial view

## Broken View

- Click on the **Broken View** option from the **View visibility** drop-down and then click on the **+** button to add the break lines. The **Drawing View** dialog box will be displayed as shown in Figure-19.

*Figure-19. Drawing View dialog box for broken view*

- Click on the straight edge of the model to define the starting point of break line; refer to Figure-20.

*Figure-20. Break line creation*

- Click to specify the end point of the break line. You are asked to specify the starting point for the second break line.
- Click to specify starting point of second break line. The break line will be created; refer to Figure-21.

*Figure-21. Second break line*

- Move the slider in the dialog box to right and select the desired **Break Line Style** from the drop-down; refer to Figure-22.

*Figure-22. Break line style options*

- After selecting the desired style, click on the **Apply** button from the dialog box to create the broken view; refer to Figure-23.

*Figure-23. Broken view created*

## Scale Options

- Click on the **Scale** option from the **Categories** list box to display the options related to scaling; refer to Figure-24.

*Figure-24. Scale options*

- Select on the **Custom scale** radio button and specify the desired value for scaling the view in the adjacent edit box.
- Click on the **Apply** button to apply the scaling.

The options for sections will be discussed later in the chapter. Now, we will continue to View States options.

## View States options

The **View States** options are generally used for assembly models, so you should have assembly model selected as Drawing Model.

- Click on the **View States** option from the **Categories** list box. The options in the dialog box will be displayed as shown in Figure-25.

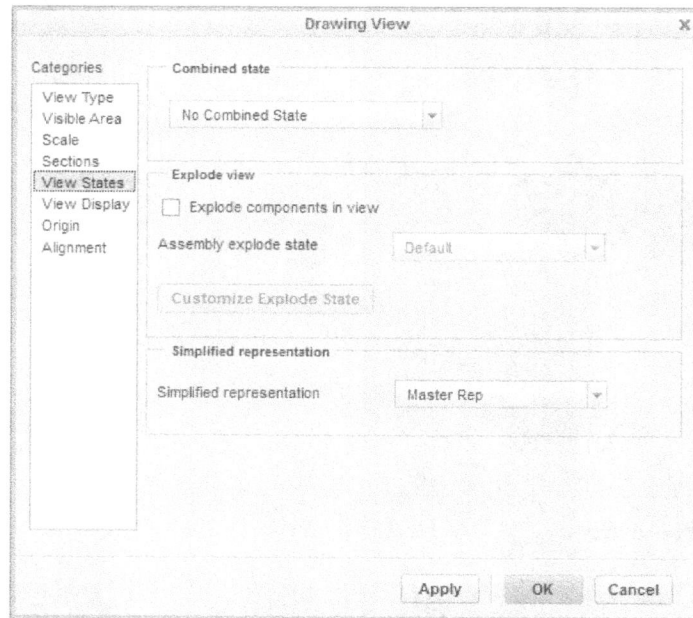

*Figure-25. Drawing View dialog box with View States options*

- Select the **Explode components in view** check box to enable the options related to exploded view of assembly model.

- Select the desired exploded state of the model from the **Assembly explode state** drop-down and click on the **Apply** button to display the exploded view; refer to Figure-26.

Figure-26. Exploded view of assembly

- Select the desired representation from the **Simplified representation** drop-down. By default, **Master Rep** is selected in the drop-down.

## View Display options

- Click on the **View Display** option from the **Categories** list box. The options in the dialog box will be displayed as shown in Figure-27.

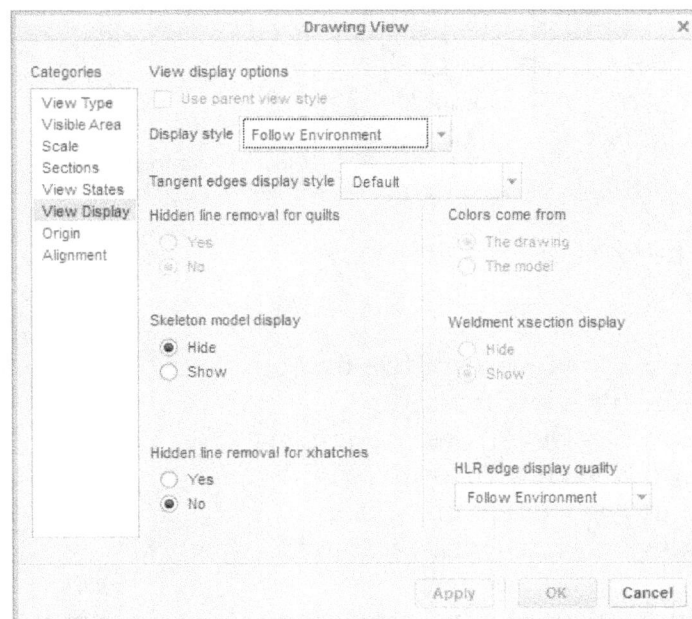

Figure-27. Drawing View dialog box with View Display options

- Click on the **Display style** drop-down and select the desired option from the drop-down to specify the display style; refer to Figure-28.

*Figure-28. Display style drop-down*

• Specify the other display style options from the dialog box and click on the **Apply** button to accept the changes.

The **Origin** and **Alignment** options in the dialog box are used to specify origin of view and alignment for the view respectively.

## PROJECTION VIEW

The **Project View** tool is used to create projections of the base view. Before we go for projection view, we need to revise the concepts of First Angle Projection and Third Angle Projection. Figure-29 shows an object with different view directions say, a, b, c, d, e, and f.

Object with various view directions

*Figure-29. Object with view directions*

Here,

1. View in the direction a = view from the front
2. View in the direction b = view from top
3. View in the direction c = view from the left
4. View in the direction d = view from the right
5. View in the direction e = view from bottom
6. View in the direction f = view from the back

In First Angle projection, these views are arranged as shown in Figure-30. In Third Angle projection, these views are arranged as shown in Figure-31.

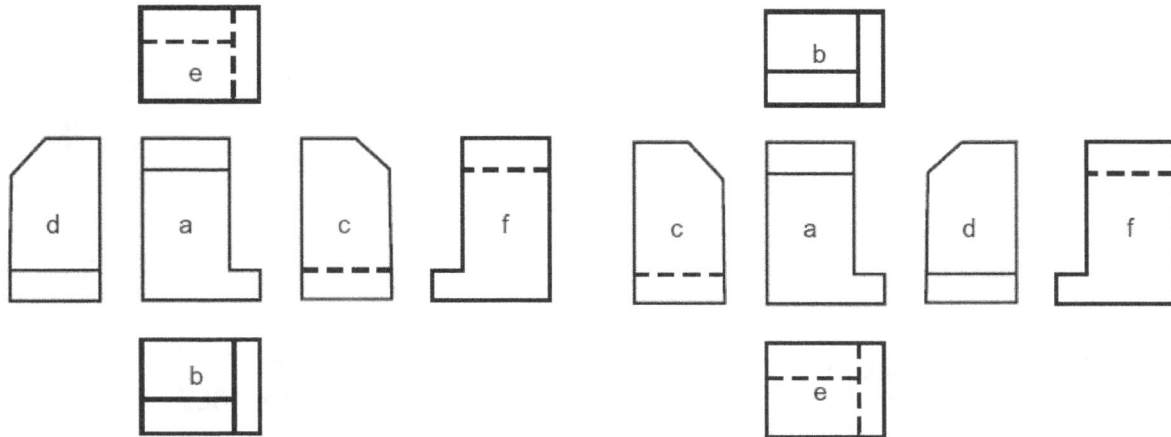

Views of an object

Views in Third Angle projection

Symbol of First Angle

Symbol of Third Angle projection

*Figure-30. Views in First Angle projection*

*Figure-31. Views in Third Angle projection*

## Setting Projection Type

In Creo Parametric, we can set the project type between First Angle and Third Angle by using the **Drawing Properties** option in the **File** Menu. The procedure is given next.

- Expand the **File** Menu and select **Drawing Properties** option from the **Prepare** cascading menu; refer to Figure-32. The **Drawing Properties** dialog box will be displayed; refer to Figure-33.

*Figure-32. Drawing Properties option*

*Figure-33. Drawing Properties dialog box*

- If you want to change the tolerance standard then click on the **change** option in the end of **Tolerance** row in the dialog box. The **TOL SETUP Menu Manager** will be displayed. Click on the **Standard** option and then select the desired tolerance standard; refer to Figure-34.
- Click on the **change** option next to **Detail Options** in the dialog box. The **Options** dialog box will be displayed as shown in Figure-35.

*Figure-34. TOL SETUP Menu Manager*

*Figure-35. Options dialog box*

- Type **projection type** in the **Option** edit box at the bottom of the dialog box. Two options will be displayed under the **Value** drop-down; refer to Figure-36.

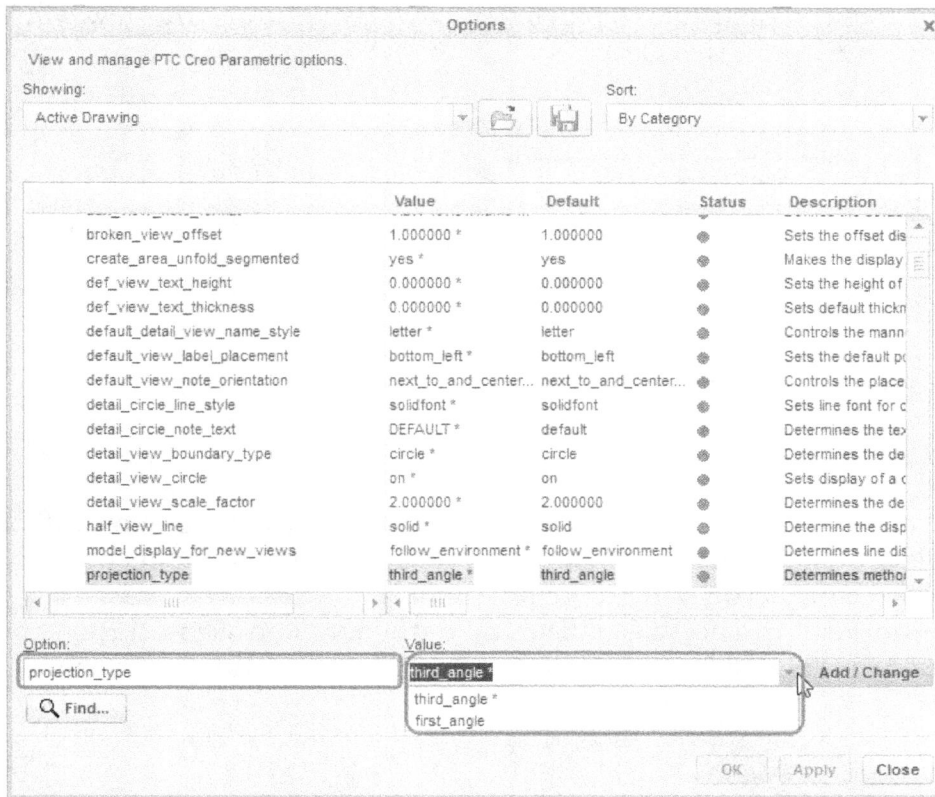

*Figure-36. Projection type options*

- Select the desired projection type from the drop-down. Click on the **Add/Change** option and then click on the **OK** button to change the projection type in the drawing. (We will continue with the default one i.e. third angle in this book.)
- Click on the **Close** button from the **Drawing Properties** dialog box to exit.

## Creating Projection Views

The steps to create projection views from the base view are given next.

- Click on the **Projection View** button from the **Model Views** panel in the **Ribbon**. Projection of the base view will get attached to the cursor; refer to Figure-37.

Base View

Preview of right side projection view

*Figure-37. Preview of projection view*

- Click at the desired distance to place the view. By default, the view will be displayed as shaded; refer to Figure-38.

*Figure-38. Shaded projection view*

- Double-click on the projection view. The **Drawing View** dialog box will be displayed; refer to Figure-39. Select the **Add projection arrows** check box if you want to display projection arrows in the drawing.
- Using the options in the **View Display** category, you can change the display of view to **Hidden**, **No Hidden**, or **Wireframe** for presentation purpose.

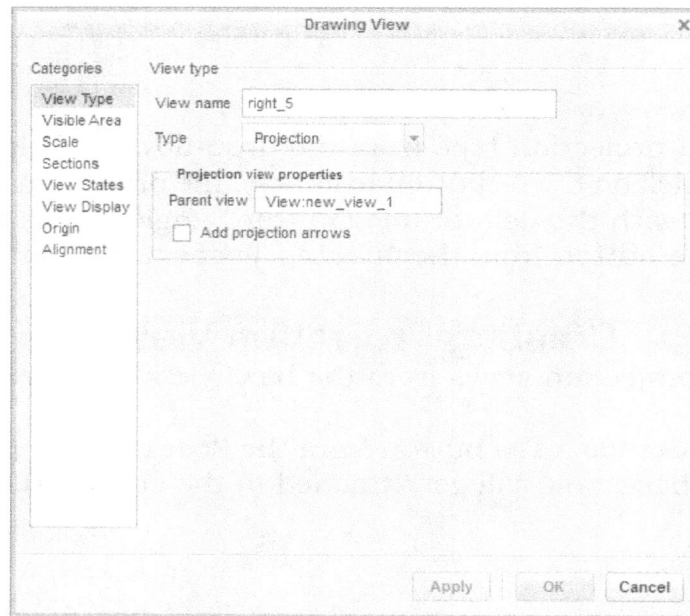

*Figure-39. Drawing View dialog box*

- Click on the **OK** button from dialog box.
- Click again on the **Projection View** button and select the base view as parent view for projection. Repeat the same procedure to place other views; refer to Figure-40.

*Figure-40. Views placed in drawing*

## Detailed View

The detailed views are used to highlight a small section of drawing which has minute details that are to be considered while manufacturing. The procedure to create detailed views is given next.

- Click on the **Detailed View** tool from the **Model Views** panel in the **Ribbon**. You are asked to specify the center point for the view.
- Click at the location on model which is center for your detailed view; refer to Figure-41. You are asked to draw a non-self-intersecting spline.

*Figure-41. Center point selection*

- Create a closed spline around the center point which is covering all the entities that you want to include in the detailed view; refer to Figure-42 and press the MMB. A circle will be drawn around the center point and you are asked to specify placement point for detailed view.

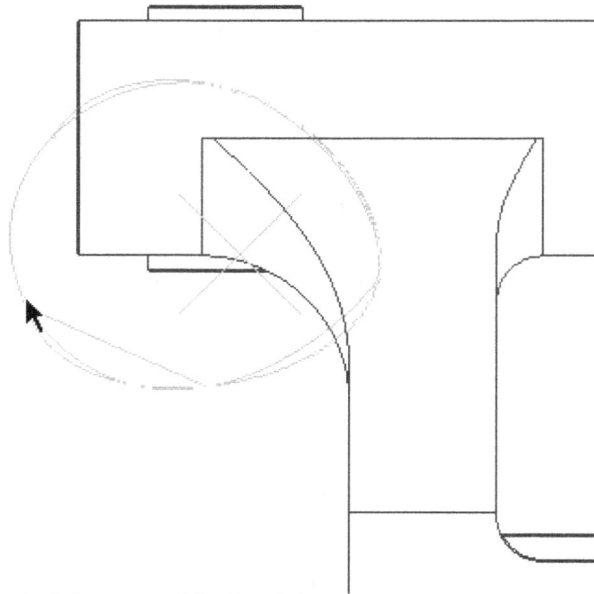

*Figure-42. Spline created for detailed view*

• Click at the desired location in the drawing sheet to place the detailed view; refer to Figure-43.

*Figure-43. Detailed view*

• Double-click on the detailed view to change the details like display style, scale and so on.

## AUXILIARY VIEW

Auxiliary views are orthographic views taken from a direction of sight other than top, front, right side, left side, bottom, or rear. Auxiliary views are often used to show inclined and oblique surface's true size. Inclined and oblique surfaces do not show true size in the standard views. The procedure to create auxiliary view is given next.

- Click on the **Auxiliary View** tool from the **Model Views** panel in the **Ribbon**. You are asked to select an edge, axis, datum plane or surface.
- Select the entity through which you want to create the auxiliary view. The view will get attached to the cursor; refer to Figure-44.

*Figure-44. Auxiliary view attached to cursor*

- Click at the desired location to place the view; refer to Figure-45.

*Figure-45. Auxiliary view placed*

## REVOLVED VIEW

The Revolved view is created to show detail of a part after revolving and sectioning a specific area of part. This type of view is used to show hidden details of part. The procedure is given next.

- Click on the **Revolved View** tool from the **Model Views** panel in the **Ribbon**. You are asked to select parent view.
- Select the view for which you want to create the revolved view. You are asked to specify center point for revolved view.
- Click at the desired location to place the view. The **XSEC CREATE Menu Manager** will be displayed along with the **Drawing View** dialog box.

- Select the **Planar** and **Single** option from the Menu Manager and click on the **OK** button. (You will learn about other options later in this chapter).
- Click on the **Done** button from the Menu Manager. You are asked to specify name for the section.
- Enter the desired name in the input box displayed. The **SETUP PLANE Menu Manager** will be displayed.
- Click on the **Make Datum** option and select the desired option from the **DATUM PLANE** area in the Menu Manager. Note that we are going to create a datum plane which will act as cutting plane for creating the cross-section.
- After selecting the desired option (which is Through in our case), select the respective reference (which is an edge in our case; refer to Figure-46.)

*Figure-46. Edge selected*

- Click on the **Done** option from the Menu Manager. Preview of the revolve view will be displayed. Note that it can be overlapping with the parent view; refer to Figure-47.

*Figure-47. Preview of revolved view*

- Click on the **OK** button from the **Drawing View** dialog box to create the view.
- Release the **Lock View Movement** toggle button from the **Document** panel in the **Ribbon** and drag the view to desired distance from the base view; refer to Figure-48.

*Figure-48. Revolved view*

## CREATING SECTION VIEWS

Section views are used to display hidden details of an object by cutting it from selected reference. To create a section view in Creo, you must have a view already present in the drawing that can be sectioned. The procedure to create section views is given next.

• Create the view that you want to be sectioned; refer to Figure-49.

Base View

View to be sectioned

*Figure-49. View created for sectioning*

• Double-click on the view to be sectioned. The **Drawing View** dialog box will be displayed.
• Click on the **Section** option from the **Categories** list box. The dialog box will be displayed as shown in Figure-50.

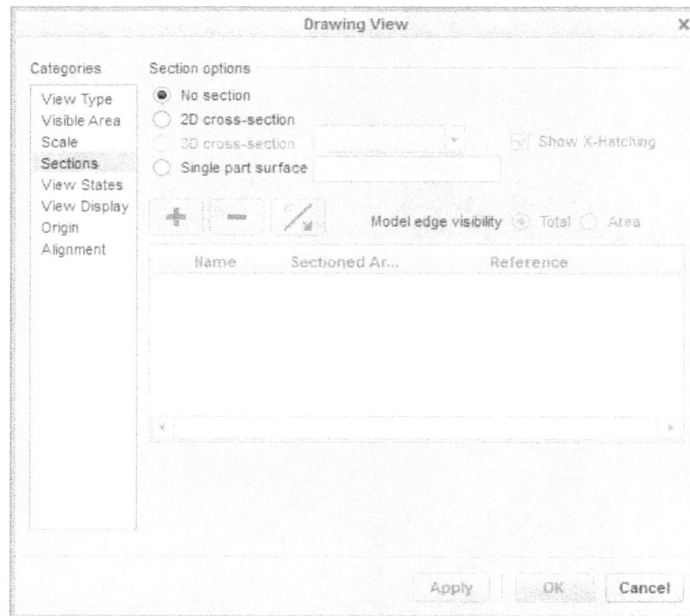

*Figure-50. Drawing View dialog box with Section options*

## 2D Cross-sectioning

Here, we will section the part in 2 Dimensional plane.

- Select the **2D cross-section** radio button and click on the **+** button from the dialog box. The **XSEC CREATE** Menu Manager will be displayed as discussed earlier. We have discussed planar section options earlier. Now, we will start with Offset section.
- Select the **Offset**, **Both Sides**, and **Single** option from the **XSEC CREATE** Menu Manager and then click on the **Done** button. You are asked to specify the name for section.
- Specify the desired name in the Input box displayed in screen and click on the **OK** button from the Input box. A separate window with sketching options will be displayed; refer to Figure-51.

*Figure-51. Window for sketching cross-section boundary*

- Select the sketching plane; refer to Figure-52. (You can create a datum plane by using the Menu Manager displayed, if needed). You are asked to specify sketching direction and references for orienting the part.

*Figure-52. Sketching plane selected*

- Select the default options or as required. Now, we need to orient the part parallel to screen for sketching, if not oriented by default.
- Click on the **View Manager** tool from the **View** menu; refer to Figure-53. The **View Manager** dialog box will be displayed.

*Figure-53. View Manager*

- Click on the **Orient** tab. Various predefined orientations will be displayed; refer to Figure-54.

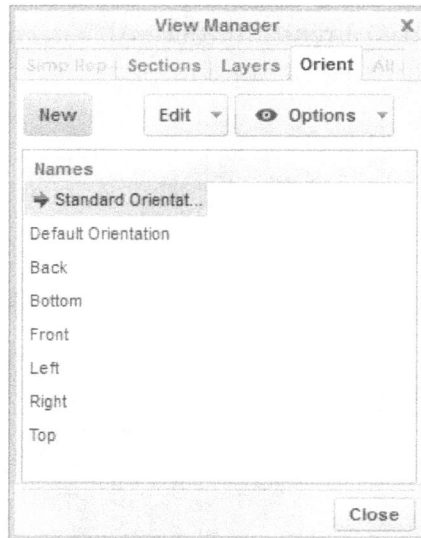

*Figure-54. View Manager with Orient tab*

- Double-click on the orientation which makes the sketching plane parallel to screen.
- Click on the **Close** button from the **View Manager** dialog box to exit.
- Select the **Line** tool from the **Sketch** menu and create a line sketch for making cross-section boundary; refer to Figure-55.

*Figure-55. Sketch drawn*

- After drawing sketch, click on the **Done** button from the **Sketch** menu. The section name will appear with a green tick mark in the **Section** options area of the **Drawing View** dialog box. The Green tick mark denotes that select section is applicable on selected view. If a red cross mark is displayed in place of the green tick mark, then either the view you have selected for sectioning is not correct or the section sketch you have created is not intersecting the view.
- Click on the **Apply** button to create the section; refer to Figure-56.

*Figure-56. Model with section view*

# ANNOTATIONS

Annotations are used to displayed geometric and dimensional tolerances of the parts. In Creo Parametric, you can apply the dimensions manually or you can import the dimensions from Part or assembly environment. First, we will discuss the method to import dimensions from Part/Assembly environment and then we will discuss the method to manually apply dimensions.

## Importing dimensions

When we create model of a part/assembly, we apply some dimensions to it in Part/ Assembly environment. These dimensions decide the shape and size of the part/ assembly. We can import these dimensions in the drawing environment by following the steps given next.

- Click on the **Show Model Annotations** button from the **Annotations** panel in the **Annotate** tab of the **Ribbon**; refer to Figure-57. The **Show Model Annotations** dialog box will be displayed; refer to Figure-58.

*Figure-57. Show Model Annotations button*

*Figure-58. Show Model Annotations dialog box*

- Select the view to which you want to apply the annotations. Dimensions will be displayed on the view; refer to Figure-59. Note that you can select multiple views by holding the CTRL key while selecting.

*Figure-59. Dimensions displayed with the views*

- One by one, click on the dimensions in the drawing that you want to keep or click on the **Select All** button to displayed all the dimensions.
- Click on the **Apply** button to display the selected dimension in the view.
- In the same way, you can import other data by using other tabs in the **Show Model Annotations** dialog box. The purpose of various tabs is shown in Figure-60.

*Figure-60. Detail of tabs*

To change any of the dimension, double-click on it and specify the desired value. On selecting the **Regenerate** button from **Quick Access Toolbar**( refer to Figure-61),the modifications are applied to the model.

*Figure-61. Regenerate button*

## Manual Dimensioning (Dimension Tool)

Like Part environment, we also have manual dimensioning tool in Drawing environment. The only drawback of using manual dimensioning in Drawing environment is that, the model do not update as per the dimensions. For example, if you have changed manual dimension of a line then its will be display as changed but the line will remain the same. The procedure to apply manual dimensions is given next.

•   Click on the **Dimension** tool from the **Annotations** panel in the **Ribbon**. The **Select Reference** dialog box will be displayed; refer to Figure-62.

*Figure-62. Select Reference dialog box*

•   Select the entities like we have done in the chapter for sketching and place the dimensions by pressing MMB.

- Select the **Tangent** button from the **Select Reference** dialog box and you will be able to dimension circles and arcs at their tangency points; refer to Figure-63.

*Figure-63. Dimensioning arc at tangency*

- Using the **Midpoint** button, you can dimension the entities with reference to their mid points; refer to Figure-64. Note that you can select multiple entities by holding the CTRL key while selecting entities.

*Figure-64. Dimensioning with reference to mid points*

- Similarly, you can use the other options in **Select Reference** dialog box.

## Ordinate Dimension

The **Ordinate Dimension** tool works in the same way as it does in Part environment. The procedure is given next.

- Click on the **Ordinate Dimension** tool from the **Ordinate Dimension** drop-down in the **Annotations** panel in the **Ribbon**; refer to Figure-65. The **Select Reference** dialog box will be displayed.

*Figure-65. Ordinate Dimension tool*

- Select the reference edge which will be set a zero reference and then select all the entities to which you want to apply the dimensions while holding the CTRL key. Preview of ordinate dimensions will be displayed; refer to Figure-66.

*Figure-66. Preview of ordinate dimensions*

- Press the MMB at desired location to place the dimensions.

## Auto Ordinate Dimension

Using this tool, you can apply the ordinate dimension to all the entities on the selected surface/surfaces. Note the if you want to apply ordinate dimension to more than one surfaces then the surfaces must be parallel to each other. The procedure to use **Auto Ordinate Dimension** tool is given next.

- Click on the **Auto Ordinate Dimension** tool from the **Ordinate Dimension** drop-down. You are asked to select one or more surfaces.

- Select the surface whose entities are to be dimensioned; refer to Figure-67.

**Surface selected**

*Figure-67. Surface selected*

- Press the MMB. You are asked to select a base line entity.
- Select an edge, curve or datum plane. Preview of ordinate dimensions will be displayed; refer to Figure-68.

**Base line selected**

*Figure-68. Ordinate dimension created*

- Click on the **Select Base Line** option from the Menu Manager displayed to select base line for other direction; refer to Figure-69.
- Press the Middle Mouse Button to apply the dimensions.

*Figure-69. Ordinate dimension with two base lines*

## GEOMETRIC TOLERANCE

Geometrical tolerance is defined as the maximum permissible overall variation of form or position of a feature.

Geometrical tolerances are used,

(i) to specify the required accuracy in controlling the form of a feature.

(ii) to ensure correct functional positioning of the feature.

Figure-70 shows a geometric symbol with meaning of each box.

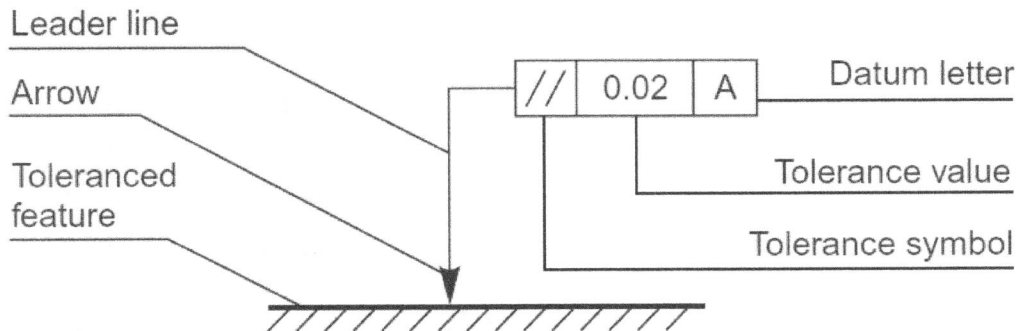

*Figure-70. Geometric Tolerance symbol*

Before we start using the tools in Creo Parametric, it is important to understand the meaning of each Tolerance Symbol.

Meaning of various geometric tolerance symbols are given in Figure-71.

| Characteristics to be toleranced | | Symbols |
|---|---|---|
| Form of single features | Straightness | ⎯ |
| | Flatness | ▱ |
| | Circularity (roundness) | ◯ |
| | Cylindricity | ⌭ |
| | Profile of any line | ⌒ |
| | Profile of any surface | ⌓ |
| Orientation of related features | Parallelism | // |
| | Perpendicularity (squareness) | ⊥ |
| | Angularity | ∠ |
| Position of related features | Position | ⊕ |
| | Concentricity and coaxiality | ◎ |
| | Symmetry | ≡ |
| | Run-out | ↗ |

*Figure-71. Meaning of geometric Tolerance symbol*

Figure-72 and Figure-73 shows the use of geometric tolerances in real-world.

*Figure-72. Use of geometric tolerance 1*

| 4. Cylindricity tolerance | 8. Concentricity and coaxiality tolerance |
|---|---|
| ⌀ 0.1 | Ⓐ ◎ φ 0.01 A |
| 5. Parallelism tolerance | 9. Symmetry tolerance |
| // 0.01 D    D | A  ≡ 0.08 A |
| 6. Perpendicularity tolerance | 10. Radial run-out |
| ⊥ 0.08 A    A | ↗ 0.1 D    D |
| 7. Angularity tolerance | 11. Axial run-out |
| ∠ 0.08 A    40°    A | ↗ 0.1 D    D |

*Figure-73. Use of geometric tolerance 2*

Note that in applying most of the Geometrical tolerances, you need to define a datum plane like in Perpendicularity, Parallelism and so on. So, before we place geometric tolerance symbol, we need to understand how to place the datum plane symbol.

## Placing Datum symbol

- Click on the **Datum Feature Symbol** tool from the **Annotations** panel in the **Ribbon**; refer to Figure-74. You will be asked to select entity to which datum symbol is to be applied.

*Figure-74. Datum Feature Symbol tool*

- Click on the edge of model, plane, axis, or dimension in drawing view. The datum symbol will get attached to the cursor. Press the MMB at desired location. The symbol will be placed and **Datum Feature** contextual tab will be displayed in the **Ribbon**; refer to Figure-75.

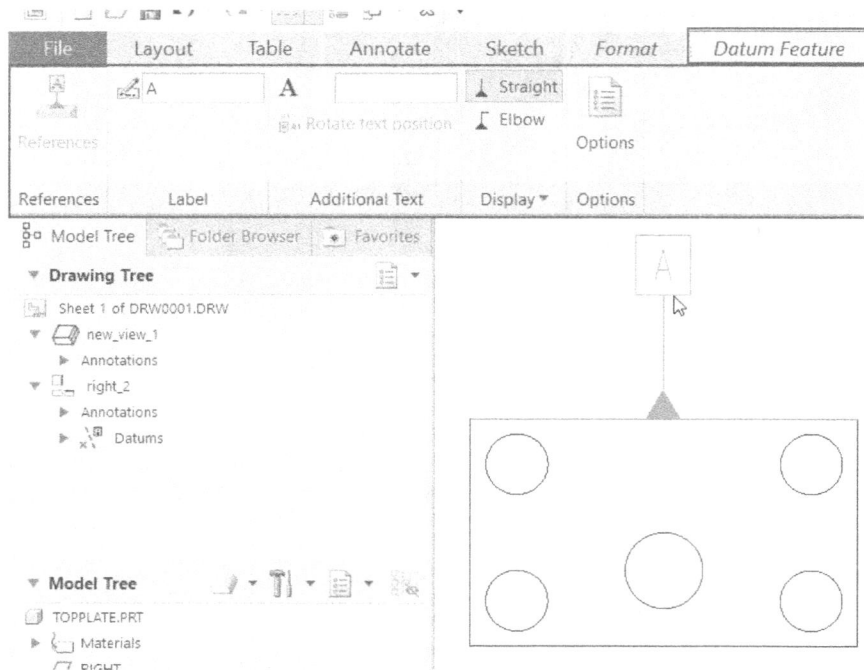

*Figure-75. Datum Feature contextual tab*

- Specify desired label (A,B,C, etc.) for datum feature in the **Label** edit box.
- If you want to specify additional text with datum feature then click in the **A** edit box next to **Label** edit box in the **Ribbon** and enter desired text.
- If you want to modify the symbol from straight to elbow type then select the **Elbow** button. If you want to display the symbol as per **ASME Y14.5m-1982** standard then select the **Display per ASME Y14.5m-1982** check box from the expanded **Display** panel of the **Ribbon**.
- Drag the datum symbol at desired location and click in the blank area of the drawing to exit the contextual tab.

## Inserting Geometric Tolerance

- Click on the **Geometric Tolerance** button from the **Annotations** panel in the **Ribbon**; refer to Figure-76. The symbol will get attached to the cursor.

*Figure-76. Geometric Tolerance button*

- Click on the edge/face/curve/dimension to which you want to apply geometric tolerance and press the MMB to specify location. The **Geometric Tolerance** contextual tab will be displayed in the **Ribbon**; refer to Figure-77.

*Figure-77. Geometric Tolerance contextual tab*

- Click on the **Geometric Characteristic** button from the **Symbol** panel in the contextual tab and select the desired GTOL (Geometric Tolerance) symbol; refer to Figure-78.

*Figure-78. Geometric Characteristic drop-down*

- Specify the desired value of geometric tolerance in the **Tolerance value** edit box in the contextual tab.
- Click on the **Select datum reference** button next to **Primary datum reference** edit box. The Select dialog box will be displayed and you will be asked to select a datum reference; refer to Figure-79.

*Figure-79. Select dialog box*

- Select the desired datum reference symbol from the drawing area and click on the **OK** button from the **Select** dialog box. If you have selected correct datum reference as per the GTOL symbol selected then the datum reference symbol will be displayed in green color in the **Primary datum reference** edit box; refer to Figure-80. If the symbol selected is not as per the datum reference then a red line will be displayed below the symbol name in the **Primary datum reference** edit box; refer to Figure-80.

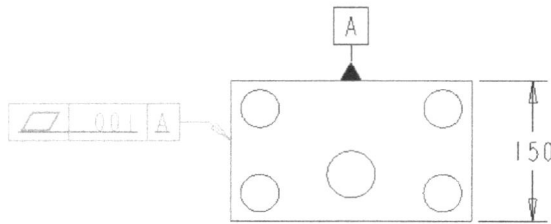

When Datum reference selected is not applicable for GTOL symbol selected

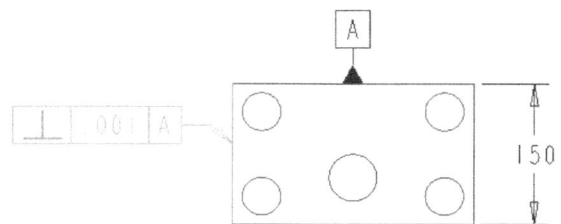

When Datum reference selected is according to GTOL symbol selected

*Figure-80. GTOL symbol and datum reference selection*

- Similarly, you can specify Secondary and Tertiary datum references using the respective options in the **Tolerance & Datum** panel of the **Geometric Tolerance** contextual tab.

## Composite Frame

- If you want to create a composite tolerance frame then click on the **Composite Frame** button from the **Tolerances & Datum** panel of the contextual tab. The **Composite Frame** box will be displayed; refer to Figure-81.

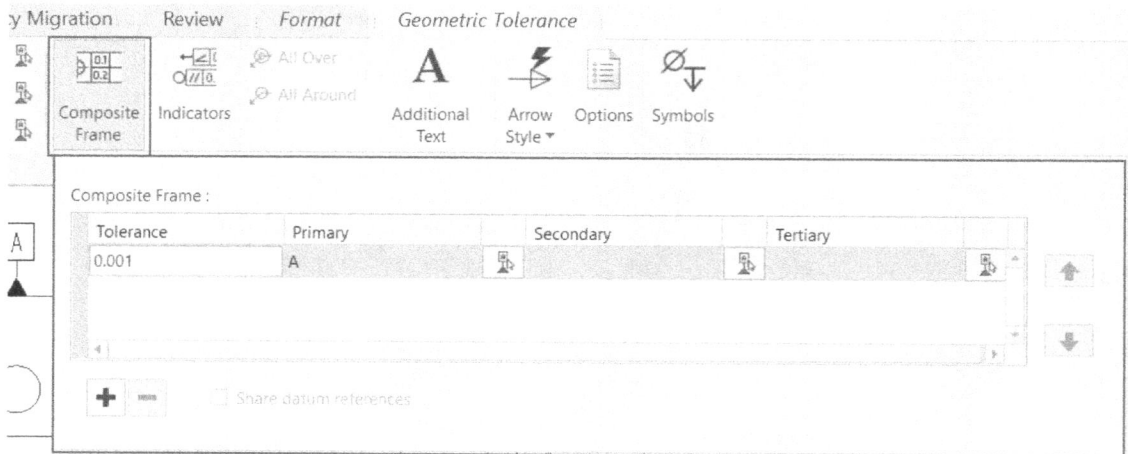

*Figure-81. Composite Frame box*

- Click on the **+** button at the bottom in the box. A new row will be added to the Geometric Tolerance frame.

- Specify the desired tolerance value and select the related datum reference; refer to Figure-82. Click on the **Composite Frame** tool again to close the box.

*Figure-82. Composite frame created*

## Indicators Frame

- Click on the **Indicators** tool from the **Indicators** panel in the **Geometric Tolerance** tab of the **Ribbon**. The **Indicators Frame** box will be displayed; refer to Figure-83.

*Figure-83. Indicators Frame box*

- Click on the **+** button and specify the desired parameters in the table.
- Click again on the **Indicators** tool to close the box.

## Additional Text

You may need to specify material conditions and other parameters along with the Geometric Tolerance frame. You can specify these parameters by using the Additional Text tool.

- Click on the **Additional Text** tool from the **Additional Text** panel in the **Geometric Tolerance** contextual tab. The input box will be displayed; refer to Figure-84.

*Figure-84. Additional Text box*

- Click in the box around the frame in which you want to specify additional text and enter the desired text in it.
- Click on the **Additional Text** tool again to close the box.

Click in the blank area of drawing to exit the contextual tab. If you want to modify a tolerance frame then click on it, the **Geometric Tolerance** contextual tab will display again.

## SURFACE FINISH SYMBOL

As the name suggests, this symbol is used to define the surface finish of face of a part. The procedure to add surface finish symbol is given next.

- Click on the **Surface Finish** tool from the **Annotations** panel of the **Ribbon**. The **Surface Finish** dialog box will be displayed; refer to Figure-85.

*Figure-85. Surface Finish dialog box*

- Click on the **Browse** button in the dialog box. The **Open** dialog box will be displayed; refer to Figure-86.

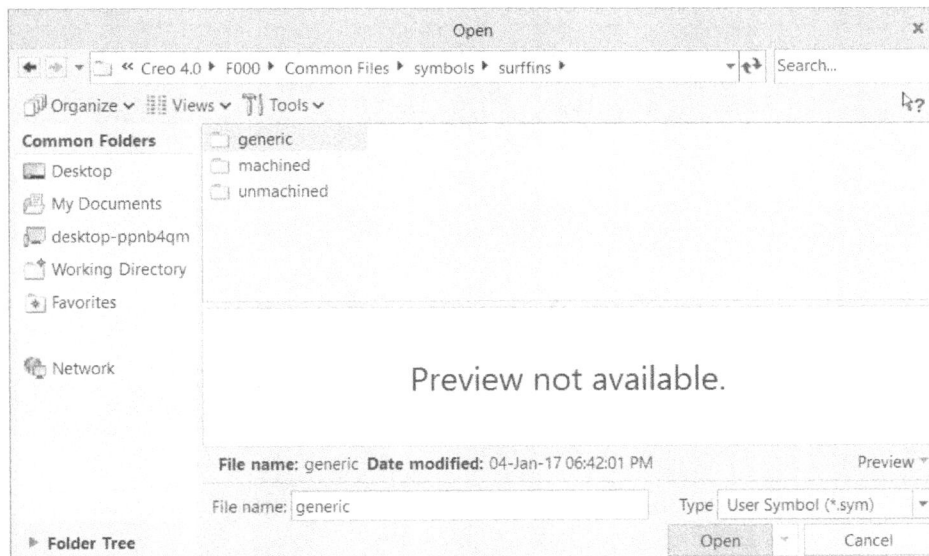

*Figure-86. Open dialog box with surface finish symbol*

- There are three categories available: Generic, machined, unmachined. Double-click on the desired folder and select the Standard symbol in the folder.
- Click on the **Open** button from the dialog box. The symbol will get attached to cursor refer to Figure-87.

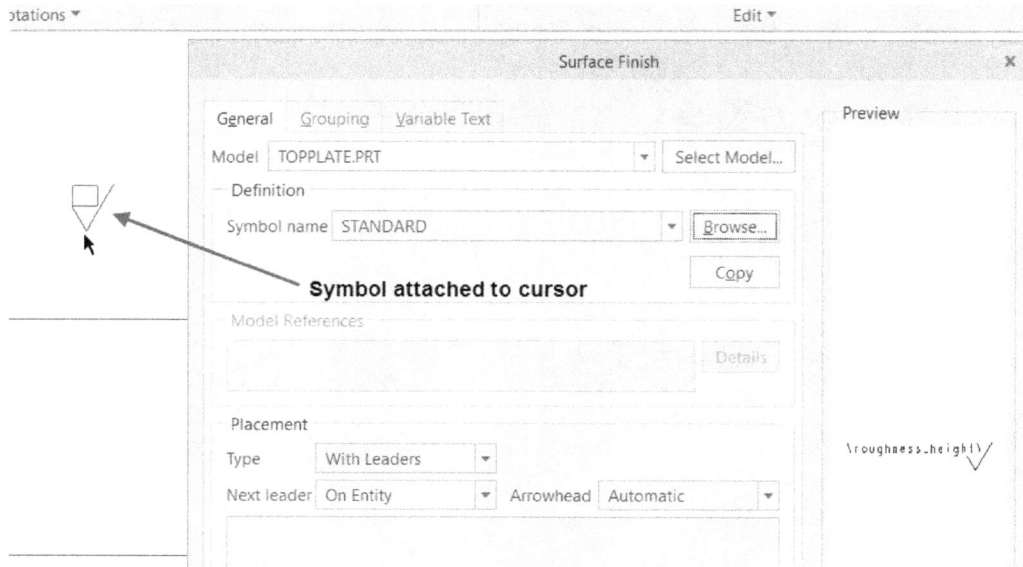
*Figure-87. Surface Finish dialog box with symbol*

- Click on the **Type** drop-down in the **Placement** area and select the desired option. Like, if you want to place the symbol with a leader then select the **With Leaders** option.
- Place the symbol by selecting the entity corresponding to option selected in **Type** drop-down; refer to Figure-88.

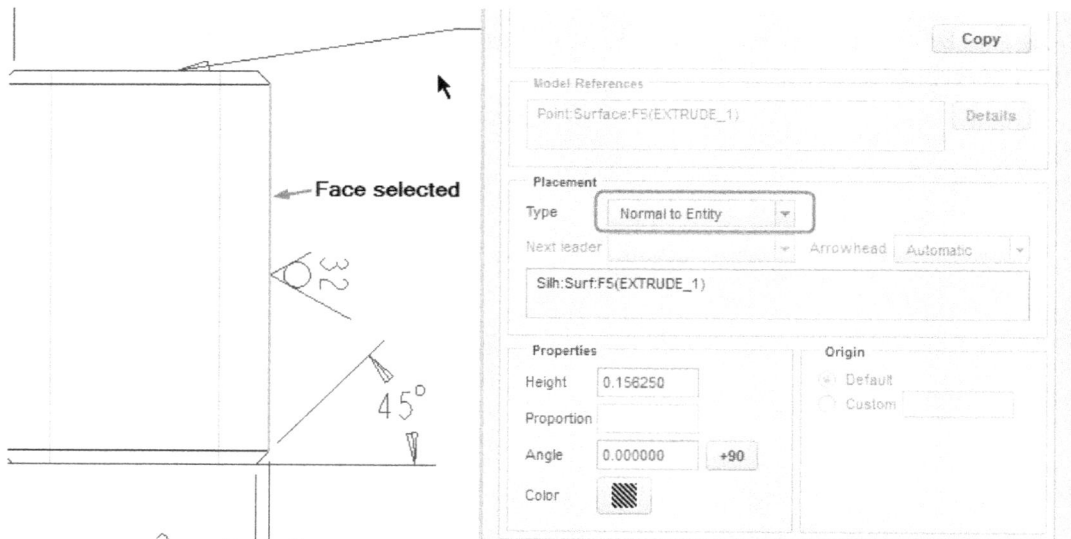
*Figure-88. Face selected for placing surface finish symbol*

- Set the desired height and angle for the symbol by using options in the **Properties** area of the dialog box.
- Click on the **Variable Text** tab and specify the value of roughness in the edit box; Figure-89.

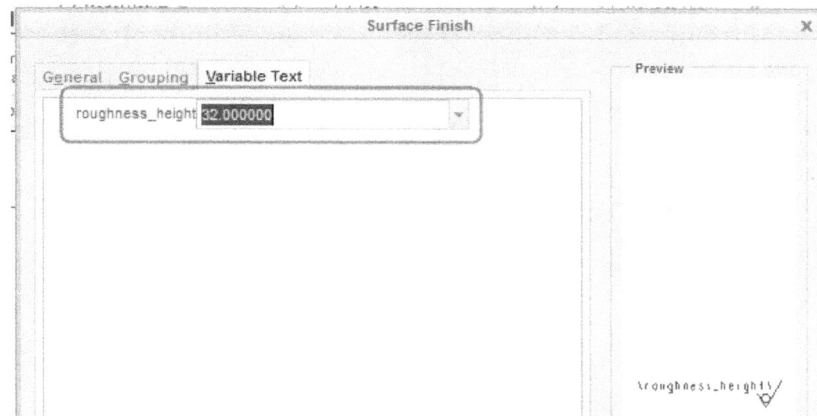

*Figure-89. Surface Finish dialog box with Variable Text tab selected*

- Press the Middle Mouse Button to place the surface finish symbol. Click on the **OK** button from the dialog box to exit.

## NOTES

Notes are used to give information which is not defined till now in the drawing with the help of dimensions and symbols. The tools to create notes are given in the **Note** drop-down in the **Annotations** panel of the **Ribbon**; refer to Figure-90. Here, we will discuss the **Unattached Note** tool and **Leader Note** tool. The other tools for creating note work in the same way.

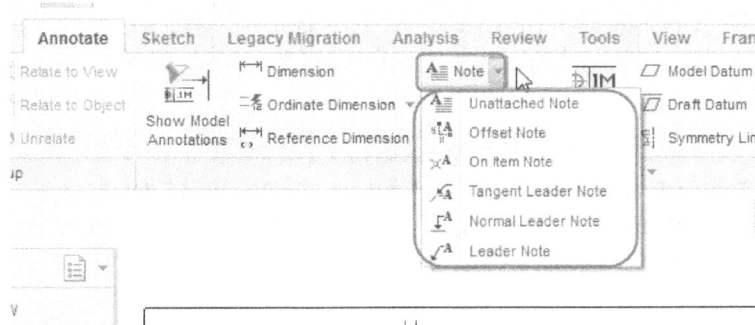

*Figure-90. Note drop-down*

## Unattached Note

- Click on the **Unattached Note** tool from the **Note** drop-down. A text box will get attached to cursor and the **Select Point** dialog box will be displayed.
- Click at the desired location to define placement of note. You are asked to type the desired text. Also, the **Format** contextual tab will be displayed; refer to Figure-91.

*Figure-91. Format contextual tab*

- Write the desire note and apply formatting and symbols by using the options in the **Format** contextual tab.
- Press the Middle Mouse Button to create the note.

## Leader Note

- Click on the **Leader Note** tool from the **Note** drop-down. A text box with leader will get attached to cursor and the **Select Reference** dialog box will be displayed.
- Click at the desired location to specify the starting point of the leader. You are asked to specify the end point of the leader.
- Press the middle mouse button (MMB) to specify the end point of leader. Rest of the procedure is same as discussed for **Unattached Note** tool.

## MODIFYING DIMENSION STYLE

- Select the **Annotation** option from the **Selection Filter** drop-down at the bottom right corner of the interface and then select all the dimensions for changing their dimension style; refer to Figure-92. The **Dimension** contextual tab will be displayed in the **Ribbon**.

*Figure-92. Dimensions selected*

- Click on the **Display** tool from the contextual tab. The options will be displayed as shown in Figure-93.

*Figure-93. Display options*

- Set the desired parameters, like change the arrow style using the **Arrow Style** drop-down or change the orientation of text using the **Text Orientation** drop-down. You can set the selected dimensions as inspection dimension using the **Inspection** check box in the box.
- If you want to change the text font or other text styles then select all dimensions and click on the **Text Style** tool from the **Annotate** tab of the **Ribbon**. The **Text Style** dialog box will be displayed; refer to Figure-94.

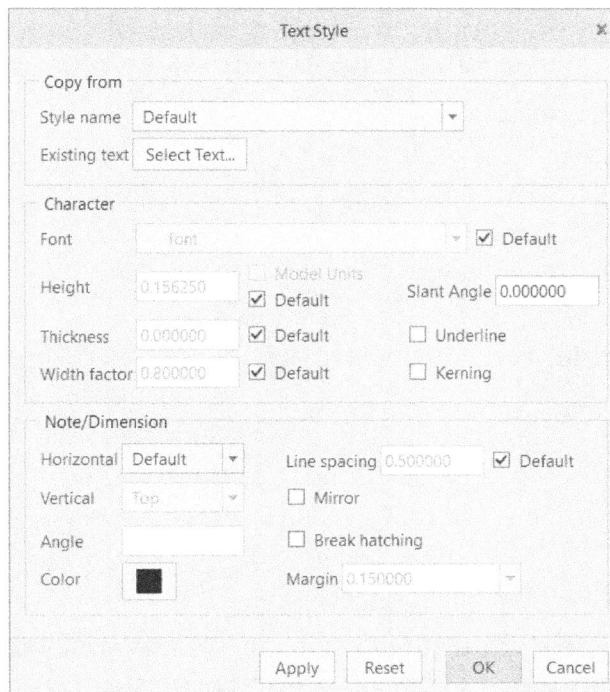

*Figure-94. Text Style dialog box*

- Clear the **Default** check box and select the desired font from the **Font** drop-down.
- Specify the other parameters related to text using the options in the dialog box and click on the **OK** button to apply the changes.

## CREATING TABLES AND BILL OF MATERIALS (BOM)

Tables are used to tabulate the data required for various purposes. For example, you can tabulate the data related to team developing the part in the drawing. The procedure to create a table in Creo Parametric is given next.

- Click on the **Table** drop-down in the **Table** panel of **Table** tab in the **Ribbon**. The options related to table insertion are displayed; refer to Figure-95.

*Figure-95. Table drop-down*

## Inserting Table dynamically

- Hover the cursor over the boxes to define the number of row and columns of table. Click on the boxes when you get desired number of rows and columns. Preview of table attached to cursor will be displayed; refer to Figure-96.

*Figure-96. Preview of table attached*

- Click to place the table; refer to Figure-97.

*Figure-97. Table placed*

- To enter data in any field, double click on it. (Make sure you have **General** option selected in the **Selection Filter** drop-down at the bottom-right corner of the interface.)

- You can perform the operations like merging cells, unmerging cells and so on by using the tools in the **Rows & Columns** panel in the **Ribbon**; refer to Figure-98.

*Figure-98. Rows and Columns panel*

- To merge the cells, select two or more cells while holding the **CTRL** key and click on the **Merge Cells** tool.
- To adjust the height and width of a cell, click on the cell and then select the **Height and Width** tool from the **Rows & Columns** panel. The **Height and Width** dialog box will be displayed; refer to Figure-99.

*Figure-99. Height and Width dialog box*

- By default, the **Automatic height adjustment** check box is selected. So, the height of column is automatically adjusted according to the data entered in the cell. To define a specific height, clear the check box and specify the value in the **Height (drawing units)** or **Height (characters)**edit box. Similarly, you can change the width of the cell.

## Inserting Table by dialog box

- Click on the **Insert Table** tool from the **Table** drop-down in the **Table** panel of **Table** tab in the **Ribbon**. The **Insert Table** dialog box will be displayed; refer to Figure-100.
- Specify the number of columns and rows of the table by using the spinners in the **Table Size** area of the dialog box.
- Specify height and width of cells by using the related edit boxes and click on the **OK** button from the dialog box. The **Select Point** dialog box will be displayed and you will be prompted to specify the insertion point for the table.
- Click in the drawing to place the table. Rest of the procedure is same as discussed earlier.

*Figure-100. Insert Table dialog box*

## Quick Tables

There are various predefined templates of tables available in the **Quick Tables** cascading menu in the **Table** drop-down; refer to Figure-101. If your desired table matches with the templates then click on the desired template and place it at desired location. Rest of the procedure is same as discussed earlier.

*Figure-101. Quick Tables*

## Bill of Materials (BOM)

Bill of Material is used to tabulate the specifications of components used in the current assembly drawing. In Creo Parametric, we have a standard method to insert BOM. We do not need to enter each component description manually in Creo. The procedure to create Bill of Materials is given next.

- Insert a table with desired number of columns and three rows. (Generally in BOMs, we have three columns: one for serial number, second for name of component, and third for quantity. Also, in First Row we type name of table i.e. Bill of Materials; in second row we type description of columns i.e. Sr. No., Name of component etc. Note that you need to merge the cells in first row to write description of table; refer to Figure-102.)

| Bill of Material | | |
|---|---|---|
| Sr. No. | Name of Component | Quantity |
| | | |

Figure-102. Bill of material table template

- Click on the **Repeat Region** tool from the **Data** panel in the **Ribbon**; refer to Figure-103. The **TBL REGIONS** Menu Manager will be displayed; refer to Figure-104.

Figure-103. Repeat Region tool

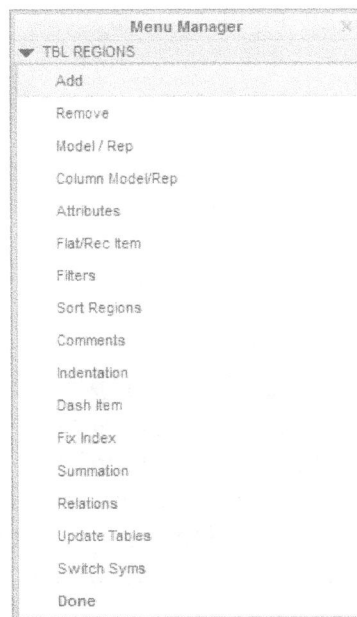

Figure-104. TBL REGIONS
Menu Manager

- Click on the **Add** option from the Menu Manager and select first and last cell of the third row in the table; refer to Figure-105.

*Figure-105. Cells selected for repeat regions*

- Press the Middle Mouse Button (MMB) once.
- Click on the **Attributes** option from the **TBL REGIONS** Menu Manager and select a cell we have added in the regions. The attribute options will be displayed in the Menu Manager; refer to Figure-106. ( If you find that you have selected only one cell and all the three cells in same row get selected! Don't worry they have become part of a region now).

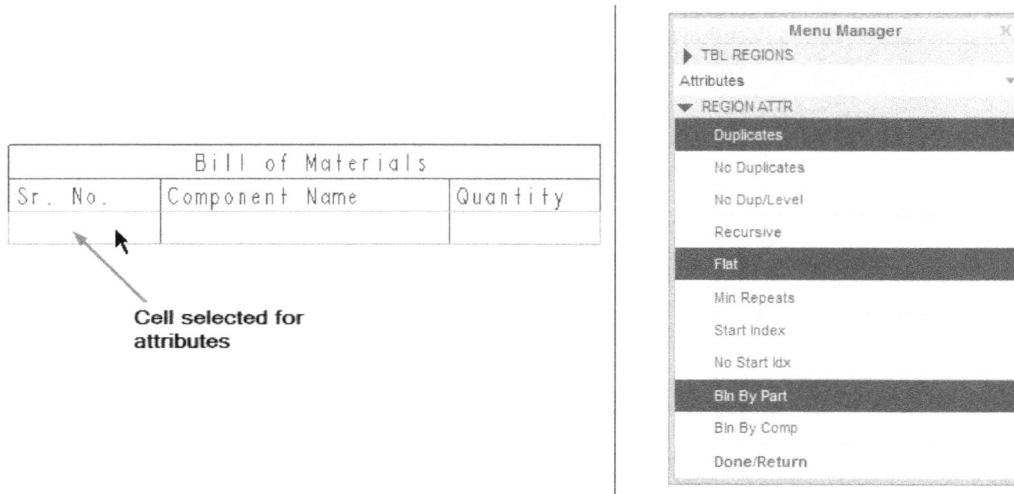

*Figure-106. Region with Attribute options*

- Select the **No Duplicates** option from the Menu Manager and press the MMB thrice to exit the Menu Manager.

## Filling Repeat Region Parameters in Table

- Double-click on the first cell of the region created (i.e. under Sr. No. column). The **Report Symbol** dialog box will be displayed; refer to Figure-107. In this dialog box, you can see various categories in their short form like, asm stands for assembly, dgm stands for diagram and so on. Details of each category and its symbols is given later in this chapter. Now, we will continue with the most basic BOM.

*Figure-107. Report Symbol dialog box*

- Select the **rpt** option and then **index** from the **Report Symbol** dialog box.
- Double-click on the middle cell in the third row and select **asm** then **mbr** and then **name** from the **Report Symbol** dialog box displayed.
- Double-click on the last cell of third row in table and select **rpt** and then **qty** from the **Report Symbol** dialog box displayed.
- Click on the **Update Tables** tool from the **Data** panel in the **Table** tab of the **Ribbon** to update the Bill of Materials table. Refer to Figure-108 for generated Bill of Materials.

| Bill of Materials | | |
|---|---|---|
| Sr. No. | Component Name | Quantity |
| 1 | BOTTOM | 1 |
| 2 | GUIDE_BUSH | 4 |
| 3 | GUIDE_PILLAR | 4 |
| 4 | TOPPLATE | 1 |

*Figure-108. Bill of materials generated*

### Categories in Report Symbol

**asm** stand for assembly, so you can find all the parameters related to assembly components in it.

**dgm** stand for diagram, so you can find all the parameters related to electrical diagrams in it.

**fam** stand for family, so you can find all the parameters related to family components in it.

**harn** stand for harness, so you can find all the parameters related to pipe/wire harness components in it.

**lay** stand for layout, so you can find all the parameters related to assembly components in it.

**material** stand for material, so you can find all the parameters related to material in it like tensile strength.

**mbr** stand for member, so you can find all the parameters related to members of current wiring/piping/cabling diagram in it.

**mdl** stand for model, so you can find all the parameters related to model in it.

**mfg** stand for manufacturing, so you can find all the parameters related to manufacturing components in it.

**prs** stand for process step, so you can find all the parameters related to process planning in it.

**rpt** stand for report, so you can find all the parameters related to quantity in it like quantity of a component in assembly.

**weldasm** stand for welding assembly, so you can find all the parameters related to welding assembly components in it.

But, we do not use all the categories in our general use. Here are some commonly used report symbols:

| Parameter Name | Definition |
|---|---|
| asm.mbr.cparams.name | Lists the names of all user-defined parameters in an assembly component. This parameter is defined for the parts making up the connector outside of the cabling environment. |
| asm.mbr.cparams.value | Lists the values of all user-defined parameters in an assembly component. This parameter is defined for the parts making up the connector outside of the cabling environment. |
| asm.mbr.cparam.User Defined | Lists the specified user-defined parameters used in an assembly component. These parameters are defined for the parts making up the Connector outside of the cabling environment. |

| asm.mbr.name | Displays the name of an assembly member. |
|---|---|
| asm.mbr.param.name | Lists the names of all user-defined parameters in an assembly member |
| mdl.param.name | Lists the names of all user-defined parameters in a model. |
| rpt.index | Displays the number assigned to each record in a repeat region |
| rpt.level | Shows the recursive depth of an item. |
| rpt.qty | Displays the quantity of an item. |

# GENERATING BALLOONS

Balloons are used to identify the components by their index numbers. Balloons are generally used in combination with Bill of Materials. The procedure to generate balloons is given next.

- Click on the **Create Balloons** drop-down from the **Balloons** panel in the **Ribbon**. The tools related to balloons will be displayed; refer to Figure-109.

*Figure-109. Create Balloons drop-down*

- Click on the **Create Balloons-All** tool from the drop-down. The balloons will be created attached to the respective component; refer to Figure-110.

*Figure-110. Balloons created*

• Click on the balloon and drag it to the desired location if it is out of sheet boundary.

# PRACTICE 1

Generate the drawing views of all the solid models we have created in Practices and Practical in the previous chapters.

# PRACTICE 2

Generate the exploded views, bill of materials and balloons from all the assembly models we have worked on till this chapter.

# PRACTICE 3

Create the model and drawing given in Figure-79.

*Figure-111. Practice 3*

# PRACTICE 4

Create the model and drawing given in Figure-80.

*Figure-112. Practice 4*

Note that the drawing given in this book are for practice purpose only.

## SELF ASSESSMENT

1. What is the full form of PMI in case of Manufacturing Technology?

2. In Drawing environment, base view is placed by using the . . . . . . . . . . . . . tool.

3. The . . . . . . . . . . . tool is used to create projections of the base view.

4. Explain First Angle and Third Angle projections with example.

5. The . . . . . . . . . . . . . . . are used to highlight a small section of drawing which has minute details that are to be considered while manufacturing.

6. The . . . . . . . . . . . . . . . . . is created to show detail of a part after revolving and sectioning a specific area of part.

7. Using this . . . . . . . . . . . . . . . tool, you can apply the ordinate dimension to all the entities on the selected surface/surfaces.

8. Which of the following symbols is for cylindricity in Geometric Tolerance box?

(a) ⌒    (b) ⌒

(c) ⌱    (d) ○

9. Which of the following symbols is for roundness in Geometric Tolerance box?

(a) ◠          (b) ◠

(c) ◇          (d) ○

10. In which category of report symbols can we find the numbers of components in assembly while generating bill of material?

(a)    **asm**          (b)    **fam**

(c)    **rpt**          (d)    **mdl**

**Answers to Self Assessment Questions :**
1. Product Manufacturing Information 2. General View 3. Project View 5. Detailed views 6. Revolved view 7. Auto Ordinate Dimension 8. (c) 9. (d) 10. (c)

# Chapter 14

# Model Based Definition (MBD) and 3D Printing

## Topics Covered

The major topics covered in this chapter are:

- *Introduction*
- *Annotation Planes*
- *Displaying Annotations*
- *Combined Views Bar*
- *GD&T Advisor*
- *Model Based Definition*
- *3D PDF Creation*
- *3D Printing Processes*
- *Part Preparation*
- *Lattice Creation*
- *Adding 3D Printer*
- *3D Printing in Creo Parametric*

# INTRODUCTION

In this chapter, we will learn about the latest trend of CAD - Model Based Definition and 3D Printing. Model Based Definition is a new way of putting Product Manufacturing Information (PMI) in CAD. Earlier, all the dimensions and tolerances were in 2D drawing but now we will learn to insert PMI directly in model. The tools to apply MBD are available in the **Annotate** tab of the **Ribbon** in Part/Assembly environment; refer to Figure-1. Various tools in this tab are discussed next.

*Figure-1. Annotate tab*

# ANNOTATION PLANES

Annotation planes are used as foundation for various annotations in Part/Assembly environment. The procedure to select an annotation plane is given next.

- Hover the cursor on a button in the **Annotation Planes Names** selection box of the **Annotation Planes** panel in the **Annotate** tab of the **Ribbon**. The plane will be displayed in the form of grid. Click on the button if you want to create annotations on this plane.
- Click on the **Active Annotation Plane** tool from the **Annotation Planes** panel in the **Ribbon**. The selected plane will become parallel to the screen.

## Annotation Plane Manager

Note that by default, **FLAT TO SCREEN**, **FRONT**, **TOP**, **RIGHT**, **BACK**, **BOTTOM**, and **LEFT** buttons are available in the **Annotation Planes Names** selection box. If you want to add more planes in the selection box, then follow the steps given next.

- Create a plane which you want to use as annotation plane and then click on the **More** button ˅ at the bottom in the **Annotation Planes** panel of the **Annotate** tab in the **Ribbon**. The **ANNOTATION PLANE MANAGER** dialog box will be displayed; refer to Figure-2.
- Click on the **New** button from the **ANNOTATION PLANE MANAGER** dialog box. The **Annotation Plane Definition** dialog box will be displayed; refer to Figure-3.

Figure-2. ANNOTATION PLANE MANAGER

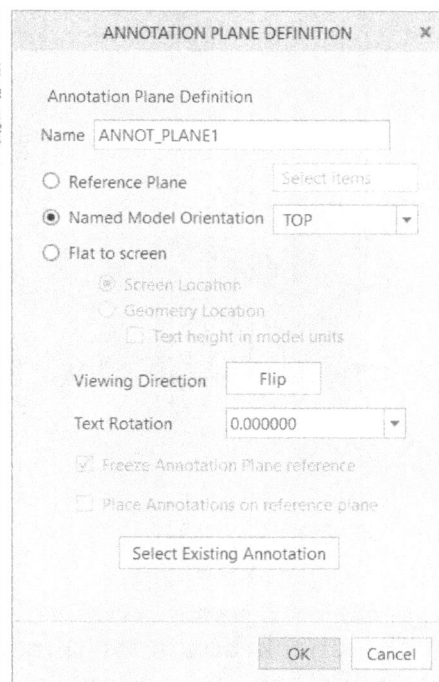

Figure-3. Annotation Plane Definition
dialog box

- Select the **Reference Plane** radio button from the dialog box. You will be asked to select a plane.
- Select the newly created plane. The plane will be displayed with grid lines.
- Specify the desired name in the **Name** edit box of the dialog box and then click on the **OK** button. The new plane will be added in the list.
- Click on the **Close** button from the **ANNOTATION PLANE MANAGER** dialog box to exit.

## SHOW ANNOTATIONS

The **Show Annotations** tool is used to display annotations assigned to the model while creating it. The procedure to use this tool is given next.

- Click on the **Show Annotations** tool from the **Manage Annotations** panel in the **Annotate** tab of the **Ribbon**. The **Show Annotations** dialog box will be displayed as discussed in previous chapter.
- Select the features of model to display their dimensions while holding the CTRL key. Respective dimension will be displayed; refer to Figure-4.

*Figure-4. Preview of annotations*

- Select the check boxes for dimensions which you want to be displayed on the model or click on the dimensions in the model.
- Click on the **OK** button from the dialog box. The annotations will be displayed on the model. Drag the annotations to place them at desired location.
- If you want to remove any annotation then select it and click on the **Erase** tool from the **Manage Annotations** panel in the **Annotate** tab of the **Ribbon**. Figure-5 shows a model after applying annotations (MBD).

*Figure-5. Model Based Definition example*

Similarly, you can use the other tools in the **Annotate** tab of the **Ribbon** which work in the same way as discussed in previous chapter on drawings.

## COMBINATION STATES

The Combination State is used to define the state of model with various annotation combinations. This option is useful when various annotations are to be displayed on the model. In that case, some dimensions can be displayed in one combination state and other dimensions can be displayed in the other combination state. The tools to create combination state are available in the **Combination States** panel of the **Ribbon**. These tools are discussed next.

## Creating New Combination State

- Click on the **New** tool from the **Combination States** panel of the **Annotate** tab in the **Ribbon**. A new combination state will be added in the **Combined Views Bar** at the bottom in the drawing area; refer to Figure-6.

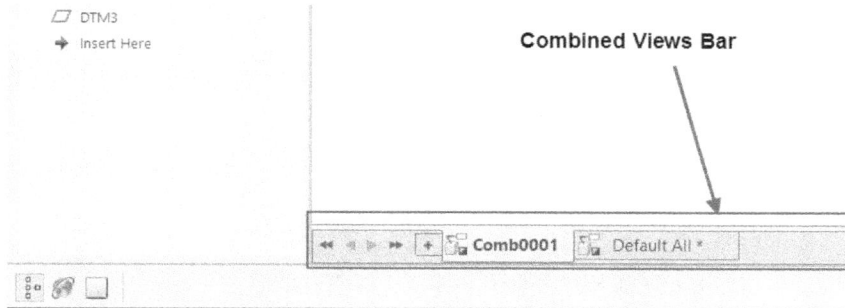

*Figure-6. Combined Views Bar*

- Apply annotations to the model and click on the **Update** button from the **Combination States** panel of the **Annotate** tab in the **Ribbon**.

## GD&T ADVISOR

The **GD&T Advisor** is an application from Sigmetrix used for applying and analyzing the model for geometric dimensioning and tolerances. The procedure to start this application is given next.

- Click on the **GD&T Advisor** tool from the **GD&T Advisor** panel in the **Applications** tab of the **Ribbon** in Part environment. The **GD&T Advisor** tab will be added in the **Ribbon**; refer to Figure-7.

*Figure-7. GD&T Advisor tab*

The tools in this tab are discussed next.

## Tolerance Feature

The **Tolerance Feature** tool is used to apply tolerance features and dimensions to the part in GD&T environment. The procedure to use this tool is given next.

- Click on the **Tolerance Feature** tool from the **Define** panel in the **GD&T Advisor** tab of the **Ribbon**. You will be asked to select one or more surfaces to apply dimensioning and tolerances.

- Select the surface(s). The **Add Feature** dialog box will be displayed with selected surfaces in the surface list; refer to Figure-8. If you select surface of a hole then other surface of hole will get automatically selected and will be identified by software. If you select a pattern surface then **Add entire pattern** check box will be selected automatically in the dialog box and the current tolerance feature will be applied to whole pattern; refer to Figure-9.

*Figure-8. Add Feature dialog box*

*Figure-9. Add Feature dialog box for pattern*

- Click on the **Accept** button from the dialog box. The **Tolerance Feature** contextual tab will be displayed in the **Ribbon** according to the object selected in the dialog box; refer to Figure-10 (**Tolerance Feature** contextual tab on selecting planar surface). If you have selected the pattern then **Tolerance Pattern** contextual tab will be displayed which will be discussed later.

*Figure-10. Tolerance Feature contextual tab*

- Select the desired tolerance symbol from the **Geometric Characteristic Symbol Selector** drop-down and specify related tolerance value in the adjacent field; refer to Figure-11.

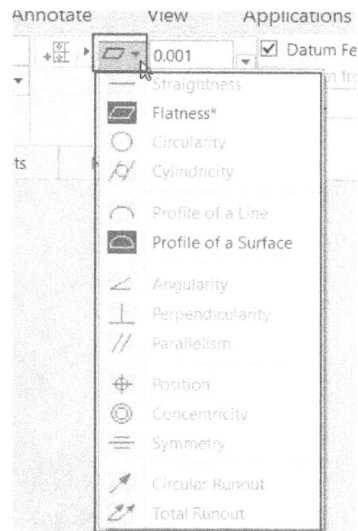

*Figure-11. Geometric Character-istic Symbol Selector drop-down*

- If you want to add a datum feature symbol to the selected face then select the **Datum Feature** check box and enter the desired label for datum feature in the **Label** edit box of the contextual tab.
- Click on the **OK** button from the contextual tab to assign the GTOL; refer to Figure-12.

*Figure-12. Datum feature preview*

- Note that if there is parallel face to the selected face at some distance then you can select the **Slab** option from the **Add Feature** dialog box; refer to Figure-13. In this way, you can also get the distance dimension between the two faces; refer to Figure-14. Click on the **Add independent segment** button to add more boxes in GTOL for creating composite GTOL box; refer to Figure-14. If you want to remove a segment from the GTOL then expand the **Segment Control Selector** flyout and select the **Remove segment** button to remove last GTOL box; refer to Figure-15.

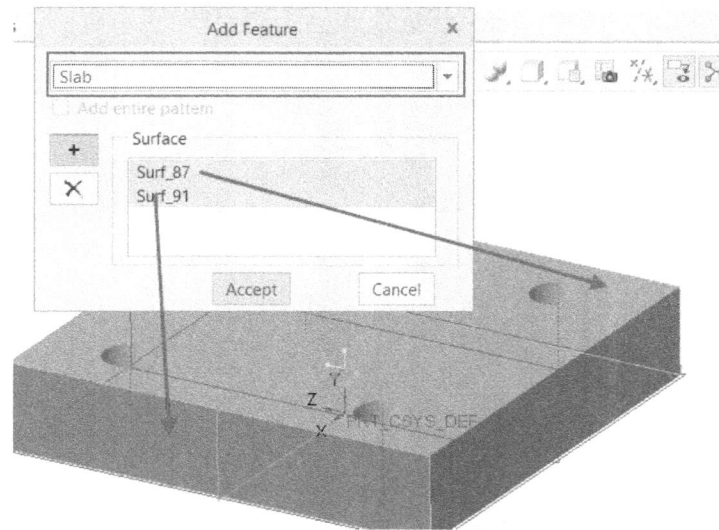

*Figure-13. Slab option in Add Feature dialog box*

*Figure-14. Slab GTOL*

*Figure-15. Remove segment button*

Similarly, you can apply GTOL to other faces of the model.

## Tolerance Pattern

The **Tolerance Pattern** tool is used to apply geometric tolerance and dimensioning to the whole pattern, like you can apply tolerance to the pattern of holes at single place. The procedure to use this tool is given next.

- Click on the **Tolerance Pattern** tool from the **Define** panel in the **GD&T Advisor** tab of the **Ribbon**. The **Add/Edit Pattern** dialog box will be displayed; refer to Figure-16.

*Figure-16. Add/Edit Pattern dialog box*

- Select one of the feature of pattern and click on the **Accept** button from the dialog box. The **Tolerance Pattern** contextual tab will be displayed; refer to Figure-17.

*Figure-17. Tolerance Pattern contextual tab*

- Specify the desired parameters in the tab as discussed earlier and click on the **OK** button to create the feature.

## Editing Properties for GD&T

The **Edit Properties** tool is used to set the standard and other parameters for applying GD&T using GD&T Advisor. The procedure to use this tool is given next.

- Click on the **Edit Properties** tool from the **Operations** panel in the **GD&T Advisor** tab of the **Ribbon**. The **Edit Part Properties** dialog box will be displayed; refer to Figure-18.
- Select the type of part for which you want to apply GD&T from the **Part Type** drop-down. Select the **Non-Rigid** check box if the part is not rigid.
- In the next options, set the desired standard for tolerancing and other parameters, and then click on the **Accept** button from the dialog box.

*Figure-18. Edit Part Properties dialog box*

The **Update** tool in **Operations** panel of the **GD&T Advisor** tab is used to update the recent changes made in GD&T of part using the tools in the **GD&T Advisor** tab of the **Ribbon**.

The **Delete All** tool in the **Operations** panel is used to delete all the GD&T data specified by using the tools in the **GD&T Advisor** tab of the **Ribbon**.

The **Show/Hide Constraint State** tool in the **Review** panel of the **GD&T Advisor** tab is used to display the constrained and unconstrained areas of the part; refer to Figure-19.

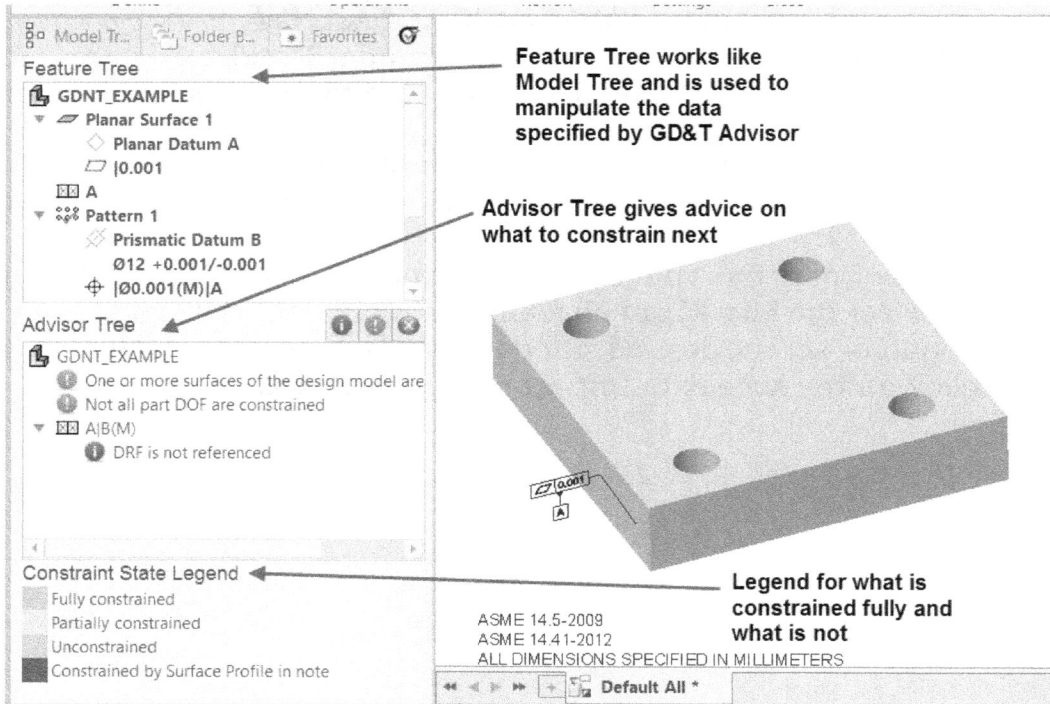

*Figure-19. Displaying constrained state of part*

Click on the **Close** button from the **GD&T Advisor** tab to exit the application. Now, you can manipulate all the GTOLs applied by **GD&T Advisor** in the same way as you have edited them earlier using the tools in **Annotate** tab of the **Ribbon**.

## 3D PDF CREATION

Although, iges and other CAD portability formats are useful data for engineers and designers but they are not useful data for marketing. For market executives, there is requirement of a format by which they can display important features of model with some dimensions. 3D PDF is one of the important format for this purpose. The procedure to create 3D PDF is given next.

*   Click on the **Save a Copy** tool from the **Save As** cascading menu in the **File** menu. The **Save a Copy** dialog box will be displayed as discussed earlier.
*   Select the **PDF U3D (*.pdf)** format from the **Type** drop-down in the dialog box and specify the desired name for the file. Click on the **OK** button from the dialog box. The **PDF U3D Export Settings** dialog box will be displayed; refer to Figure-20.

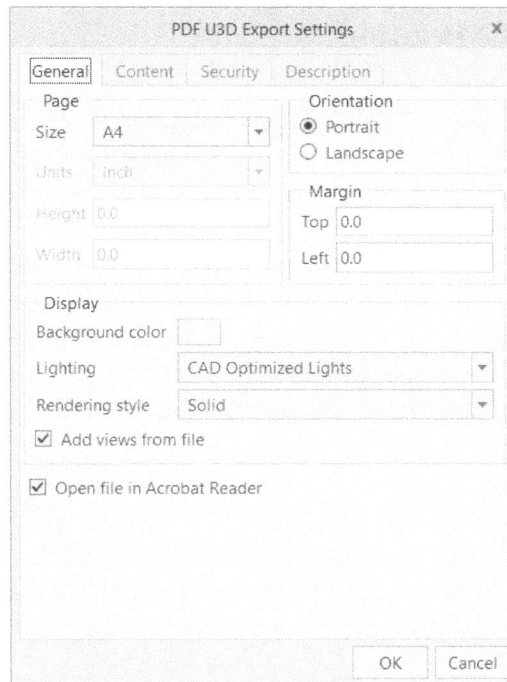

*Figure-20. PDF U3D Export Settings dialog box*

*   Select the paper size from the Size drop-down in the **Page** area of the dialog box.
*   Set the desired orientation of paper from the **Orientation** area of the dialog box.
*   Set the other parameters in the dialog box like display style.
*   Click on the **Content** tab and specify the other non-geometry features that are to be included in the 3D PDF.
*   Similarly, set the other parameters and then click on the **OK** button from the dialog box. The 3D PDF will be created and opened in the PDF reader.

## 3D PRINTING

3D Printing also known as Additive Manufacturing is not a new concept as it was developed in 1981 but since then 3D Printing technology is continuously evolving.

In early stages, the 3D printers were able to create only prototypes of objects using the polymers. But now a days, 3D printers are able to produce final products using metals, plastics and biological materials. 3D printers are being used for making artificial organs, architectural art pieces, complex design objects etc. Although, 3D Printing technique was created for manufacturing industry but now, it has found more applications in medical field.

In Creo Parametric, there is a very simple and robust mechanism for 3D printing. The procedure of 3D Printing itself is not difficult but it is important to prepare your part well for 3D printing. We will first discuss the part preparation for 3D Printing and then we will use Creo Parametric tools for performing 3D print.

## PART PREPARATION FOR 3D PRINTING

Part preparation is very important step for 3D Printing. If your part is not stable in semi molten state then it is less suitable for 3D printing. Stability of model is directly dependent on the material you are using for 3D Printing. We will know more about part preparation but before that it is important to understand different type of processes available in 3D Printing.

## 3D Printing Processes

Not all 3D printers use the same technology. There are several ways to print and all those available are additive, differing mainly in the way layers are build to create the final object.

Some methods use melting or softening material to produce the layers. Selective laser sintering (SLS) and fused deposition modeling (FDM) are the most common technologies using this way of 3D printing. Another method is when we talk about curing a photo-reactive resin with a UV laser or another similar power source one layer at a time. The most common technology using this method is called stereolithography (SLA).

In 2010, the American Society for Testing and Materials (ASTM) group "ASTM F42 – Additive Manufacturing", developed a set of standards that classify the Additive Manufacturing processes into 7 categories  according to Standard Terminology for Additive Manufacturing Technologies. These seven processes are:

1. Vat Photopolymerisation
2. Material Jetting
3. Binder Jetting
4. Material Extrusion
5. Powder Bed Fusion
6. Sheet Lamination
7. Directed Energy Deposition

Brief introduction to these processes is given next.

## Vat Photopolymerisation

A 3D printer based on the Vat Photopolymerisation method has a container filled with photopolymer resin which is then hardened with a UV light source; refer to Figure-21.

*Figure-21. 3D Printing via vat-photopolymerisation*

The most commonly used technology in this processes is Stereolithography (SLA). This technology employs a vat of liquid ultraviolet curable photopolymer resin and an ultraviolet laser to build the object's layers one at a time. For each layer, the laser beam traces a cross-section of the part pattern on the surface of the liquid resin. Exposure to the ultraviolet laser light cures and solidifies the pattern traced on the resin and joins it to the layer below.

After the pattern has been traced, the SLA's elevator platform descends by a distance equal to the thickness of a single layer, typically 0.05 mm to 0.15 mm (0.002″ to 0.006″). Then, a resin-filled blade sweeps across the cross section of the part, re-coating it with fresh material. On this new liquid surface, the subsequent layer pattern is traced, joining the previous layer. The complete three dimensional object is formed by this project. Stereolithography requires the use of supporting structures which serve to attach the part to the elevator platform and to hold the object because it floats in the basin filled with liquid resin. These are removed manually after the object is finished.

## Material Jetting

In this process, material is applied in droplets through a small diameter nozzle, similar to the way a common inkjet paper printer works, but it is applied layer-by-layer to a build platform making a 3D object and then hardened by UV light; refer to Figure-22.

*Figure-22. 3D Printing via Material-Jetting*

## Binder Jetting

With binder jetting two materials are used: powder base material and a liquid binder. In the build chamber, powder is spread in equal layers and binder is applied through jet nozzles that "glue" the powder particles in the shape of a programmed 3D object; refer to Figure-23. The finished object is "glued together" by binder remains in the container with the powder base material. After the print is finished, the remaining powder is cleaned off and used for 3D printing the next object.

*Figure-23. 3D Printing via binder-jetting*

## Material Extrusion

The most commonly used technology in this process is Fused deposition modeling (FDM). The FDM technology works using a plastic filament or metal wire which is unwound from a coil and supplying material to an extrusion nozzle which can turn the flow on and off. The nozzle is heated to melt the material and can be moved in both horizontal and vertical directions by a numerically controlled mechanism, directly controlled by a computer-aided manufacturing (CAM) software package; refer to Figure-24. The object is produced by extruding melted material to form layers as the material hardens immediately after extrusion from the nozzle. This technology is most widely used with two plastic filament material types: ABS (Acrylonitrile Butadiene Styrene) and PLA (Polylactic acid) but many other materials are available ranging in properties from wood filed, conductive, flexible etc.

*Figure-24. 3D Printing via Fused Deposition Modeling*

In the above figure:
1 – nozzle ejecting molten material (plastic),
2 – deposited material (modelled part),
3 – controlled movable table.

## Powder Bed Fusion

The most commonly used technology in this processes is Selective laser sintering (SLS). This technology uses a high power laser to fuse small particles of plastic, metal, ceramic or glass powders into a mass that has the desired three dimensional shape. The laser selectively fuses the powdered material by scanning the cross-sections (or layers) generated by the 3D modeling program on the surface of a powder bed; refer to Figure-25. After each cross-section is scanned, the powder bed is lowered by one layer thickness. Then a new layer of material is applied on top and the process is repeated until the object is completed.

All untouched powder remains as it is and becomes a support structure for the object. Therefore there is no need for any support structure which is an advantage over SLS and SLA. All unused powder can be used for the next print.

*Figure-25. 3D Printing via Selective Laser Sintering*

## Sheet Lamination

Sheet lamination involves material in sheets which is bound together with external force. Sheets can be metal, paper or a form of polymer. Metal sheets are welded together by ultrasonic welding in layers and then CNC milled into a proper shape; refer to Figure-26. Paper sheets can be used also, but they are glued by adhesive glue and cut in shape by precise blades.

## Directed Energy Deposition

This process is mostly used in the high-tech metal industry and in rapid manufacturing applications. The 3D printing apparatus is usually attached to a multi-axis robotic arm and consists of a nozzle that deposits metal powder or wire on a surface and an energy source (laser, electron beam or plasma arc) that melts it, forming a solid object; refer to Figure-27.

*Figure-26. 3D Printing via Ultrasonic Sheet Lamination.jpg*

*Figure-27. 3D Printing via Direct Energy Deposition*

Now, you know about different 3D Printing techniques so it is clear from different processes that the main work of 3D printer is to solidify material at your will, through different techniques. Now, we will learn about the points to be taken care of while preparing model for 3D printing.

## Part Preparation for 3D Printing

Various important points to remember while preparing part for 3D printing are given next.

- You should avoid holes in the areas where model is not supported. Holes can cause material to flow out in various 3D printing processes.
- Use the Solid models. It does not mean that you cannot use surface modeling tools but you should avoid surfaces in your model. Thicken your surfaces after performing modeling operations.
- Make sure that you have not left any unwanted piece inside the model enclosure after performing boolean operations.
- Shell your model after creating it. You can 3D print solid model but if a hollow box can do your work then why to waste material on solid cube. Less material means cost efficient.
- If you want to write text on your model then check the specification of your printer for possible font range.
- The textures you apply on model in Creo Parametric will not be exported for 3D printing so do not waste your time on them. Although, the color will be applied to the 3D Printed model if it is available for the printer.
- 3D printing is closer to mesh modeling than solid modeling. So, check your mesh model by exporting solid model in .stl file.
- If your file has quite a bit of text and multiple emboss/engrave features, try exporting the file as a vector. Vector files are more appropriate for extremely complex files. Try .iges or .step.
- Orientation is particularly important when it comes to 3D printing in order to determine the interior and exterior of an object. Orient your part the same way as your want it in 3D Printing.
- Don't make a multi-body model for 3D printing. Your printer may die thinking what to do when two separate bodies overlap each other!!
- A bidirectional exchange of information is supported between Creo Parametric and Stratasys & 3D Systems Printers. The Stratasys Connex Family of 3D and other models are supported.

## Lattice Creation

Lattice structures involve repetitive patterns of a particular cell shape or type. There are libraries of cell types, and the density of these cells in a design is based on the application and loading that the design experiences; refer to Figure-28. Lattice structures give more strength to the part in lesser weight. In Creo Parametric, you can convert your part to lattice structure directly which later can be used for 3D Printing. The tool to create lattice structure is available in the expanded **Engineering** panel of the **Model** tab in **Ribbon** of Part environment. The procedure to use this tool is given next.

*Figure-28. Lattice structure*

- Click on the **Lattice** tool from the expanded **Engineering** panel of the **Model** tab in the **Ribbon** of Part environment; refer to Figure-29. The **Lattice** contextual tab will be displayed; refer to Figure-30.

*Figure-29. Lattice tool*

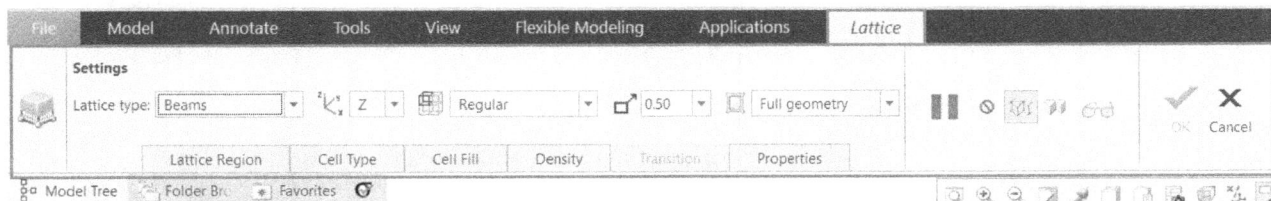

*Figure-30. Lattice contextual tab*

- Select the lattice type from the **Lattice Type** drop-down in the **Ribbon**. There are four options for lattice type; **Beam, 2.5D, Formula Driven** and **Custom**. Select the **2.5D** option if you need not control the structure of lattice but need to define the element shape. Select the **Beam** option if you want to control shape as well as structure of lattice. Figure-31 shows a component made by 2.5D lattice and Beam lattice. Note that 2.5D lattice are created faster on 3D printer and give enough strength for axial load. The Beam lattice takes time on 3D printer but is useful when there are multi-directional loads. Select the **Formula Drive** option from the drop-down if you want to use voxelized representation. Select the desired accuracy level in the Voxel size and density drop-down of **Ribbon**. You can also select the desired function from the **Function** drop-down in **Cell Type** tab of the **Ribbon**; refer to Figure-32. Selecting **Gyroid** uses sin(x) cos(y) + sin(y) cos(z) + sin(z) cos(x) = 0 formula, selecting **Primitive** uses cos(x) + cos(y) + cos(z) = 0 formula, and selecting **Diamond** uses sin(x) sin(y) sin(z) + sin(x) cos(y) cos(z) + cos(x) sin(y) cos(z) + cos(x) cos(y) sin(z) = 0 to define shape of element.

**2.5D Lattice**

**2D element extruded in Z direction**

**Beam Lattice**

**3D shaped elements**

*Figure-31. Difference between 2.5D and Beam lattice*

*Figure-32. Formula Driven lattice type*

Select the **Custom** option from the **Lattice type** drop-down if you want to use a part model earlier created in Creo Parametric as element for lattice. The options in **Lattice** contextual tab will be as shown in Figure-33. Click on the **Open Cell** button in the **Ribbon** and open the desired model to be used as element for lattice.

*Figure-33. Lattice contextual tab with Custom lattice type options*

- Select the Z direction for lattice element from the **Z-Direction** drop-down next to **Lattice Type** drop-down in the **Ribbon**.
- Using the dimensions displayed on the element preview, change the size of lattice element as required; refer to Figure-34.

*Figure-34. Lattice element preview*

- Select the multiplication method for lattice cells from the drop-down next to Z-Direction drop-down. Note that the selected method can highly affect the strength of 3D printed part.
- Similarly, select the desired representation and specify desired scale value in the **Ribbon**.
- Select the **Replace Solid with Lattice** check box from the **Lattice Region** tab of the **Ribbon** to fill the part with lattice while printing.
- If you want to create a shell of part along with lattice then select the **Create Shell** check box from the **Lattice Region** tab in the **Ribbon** and specify desired thickness of shell in the next edit box; refer to Figure-35.

*Figure-35. Lattice with shell of solid*

- To change the parameters of cells and beams of element, click on the respective tab in the **Lattice** contextual tab and set the desired parameters. Note that you can change the cell size and beam size only for 3D type lattices.
- If you want to create a variable density lattice then click on the **Density** tab in the contextual tab. The options will be displayed as shown in Figure-36.

*Figure-36. Density tab*

- Click on the **\*New set** option in the list box. A new set of variable density will be added as **Set 1**.
- Specify the distance up to which the variable density option will be applicable to lattice in the **Distance** edit box.
- Set the desired size of beam cross-section at the selected reference for variable lattice form in the **Target cross section Size** edit box and specify the rate of size change in the **Rate of Size Change** edit box of the tab.
- Specify the desired size of section up to which lattice will be created. If section size is below specified value then lattice will not be created.
- Select the **Continuous variability** check box if you want to create lattice with continuous variability.
- Click in the **References** selection box in the tab and select the faces/surfaces, edges, points etc. as starting reference for variable density lattice.

- Set the desired parameters in the **Cell Fill** tab of contextual tab in the **Ribbon**.
- Click on the **OK** button from the **Lattice** contextual tab after specifying the desired parameters to create the lattice. Save the part file in STL, OBJ, or VRML format to give it to your 3D Print Provider.

## 3D Printing in Creo Parametric

- Click on the **Prepare for 3D Printing** tool from the **Print** cascading menu in the **File** menu; refer to Figure-37. The tray assembly of current part will open in new Creo Parametric window; refer to Figure-38.

*Figure-37. Prepare for 3D Printing tool*

*Figure-38. Tray assembly of part*

- Note that the grid line is the total area of 3D Printer available for printing.
- Click on the **Assemble** button from the **Ribbon** and insert more parts if required. Note that you need to assemble the inserted parts with tray in such a way that the bottom face of part is directly on the tray; refer to Figure-39.

Placement    Properties

Set9 (User Defined )
 Coincident
   PRACTICAL_2:Surf:F7(EXTRUD!
   Surf:F6(SURFACE)

New Constraint

New Set

☑ Constraint Enabled

Constraint Type
 Coincident ▼

Offset
0.00 ▼ Flip
               Status
Partially Constrained

New part assembled with tray

Coincident

*Figure-39. Assembling part with tray*

- After inserting the parts, click on the **Arrange on Tray** button from the **Ribbon**. The parts will be placed automatically at suitable location on the table.
- Select the part and create pattern of part using the **Pattern** tool from the **Ribbon** as discussed in previous chapters; refer to Figure-40.

*Figure-40. Pattern of part on tray*

- Click on the **Arrange on Tray** tool or **Nesting** tool from the **Ribbon** to place the parts on tray automatically.
- Check the minimum thickness and gap your 3D Printer can produce from the product manual of printer. Click on the **Validate Thin Walls** tool from the **Printability Validation** drop-down in the **Analysis** panel of the **Ribbon**. The **Measure: Verify Thin Walls** dialog box will be displayed; refer to Figure-41.

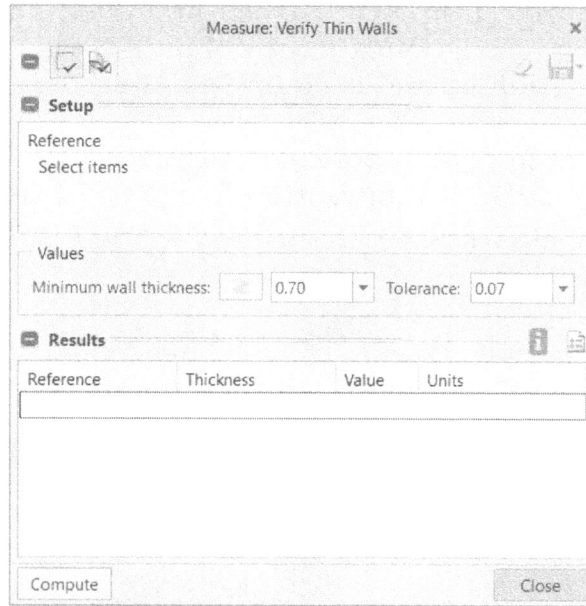

*Figure-41. Measure Verify Thin Walls dialog box*

- Select the part and specify the minimum thickness to be checked in the part. Click on the **Compute** button at the bottom in the dialog box. Increase the thickness of part by using the Flexible modeling tools if required.
- In the same way, you can perform the **Validate Narrow Gaps** analysis.
- Click on the **Generate Support** tool from the **Operations** panel in the **Ribbon** to generate support automatically.
- Click on the **Edit Support Parameters** tool from the **Generate Support** drop-down in the **Ribbon**. The **Support Structure Profiles** dialog box will be displayed; refer to Figure-42.

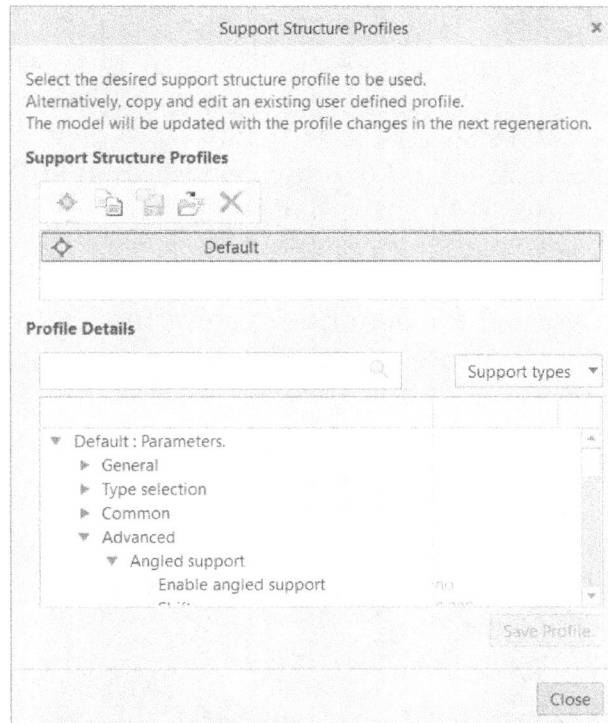

*Figure-42. Support Structure Profiles dialog box*

- Select the **Default** profile from the list box and click on the **Copy the Profile** button from the dialog box. Duplicate copy of profile will be created which can be edited. Set the desired values and click on **Set the current profile** button from the dialog box.
- Close the dialog box to apply support parameters.
- Click on the **Preview 3D Printing** button from the **Finish** panel in the **Ribbon**. The **3D Print** tab will be displayed in the **Ribbon** with preview of 3D printed parts; refer to Figure-43.

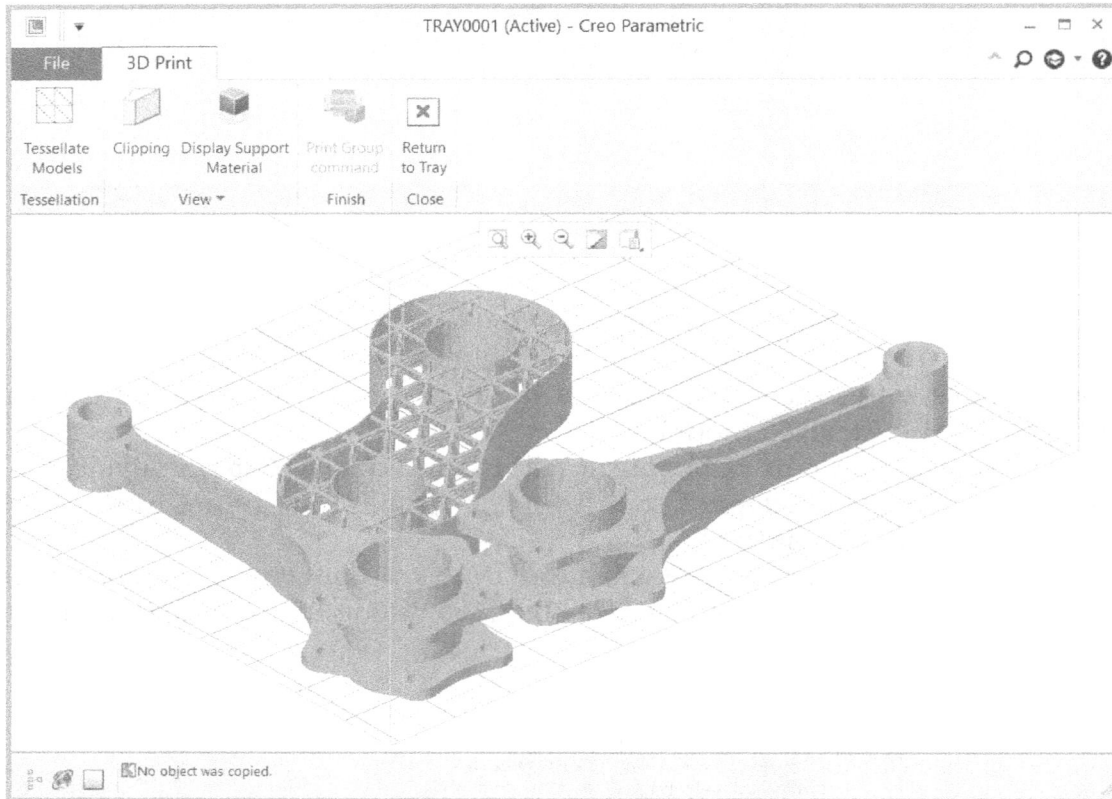

*Figure–43. 3D Print tab in Ribbon with preview*

- If you have a 3D Printer connected to your system then **Print** tool will be available in the **Finish** panel of the **3D Print** tab of the **Ribbon**. Click on the tool to directly 3D print the parts.

Till this point, we have worked for generic printers but in Creo Parametric there is a direct cost calculation and printing option if you select the **i.materialise** option from the **Printer** drop-down in the **Printers** panel of the **Tray** tab in the **Ribbon**; refer to Figure-44.

*Figure–44. Printer drop-down*

## Adding 3D Printer to Creo Parametric

- Click on the **Printers List** tool from the **Printer** drop-down in the **Printers** panel of the **Tray** tab in the **Ribbon**. The **3D Printers** dialog box will be displayed; refer to Figure-45.

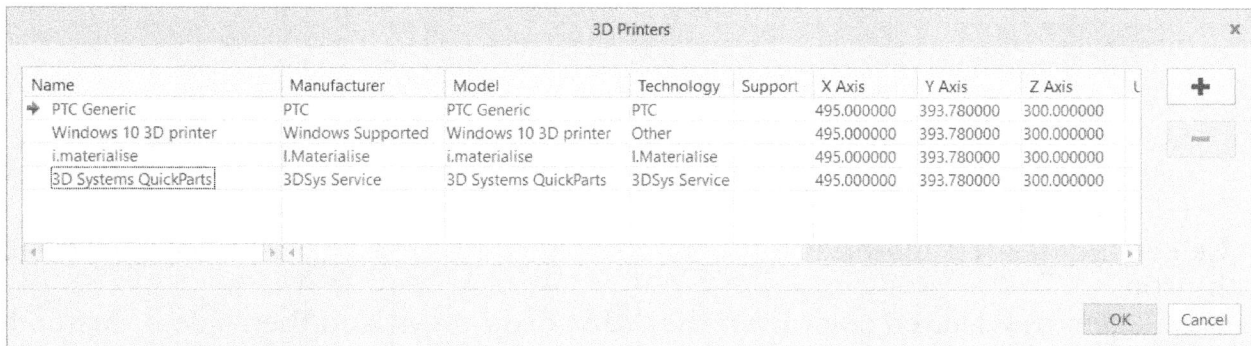

| Name | Manufacturer | Model | Technology | Support | X Axis | Y Axis | Z Axis | L |
|------|-------------|-------|-----------|---------|--------|--------|--------|---|
| → PTC Generic | PTC | PTC Generic | PTC | | 495.000000 | 393.780000 | 300.000000 | |
| Windows 10 3D printer | Windows Supported | Windows 10 3D printer | Other | | 495.000000 | 393.780000 | 300.000000 | |
| i.materialise | I.Materialise | i.materialise | I.Materialise | | 495.000000 | 393.780000 | 300.000000 | |
| 3D Systems QuickParts | 3DSys Service | 3D Systems QuickParts | 3DSys Service | | 495.000000 | 393.780000 | 300.000000 | |

*Figure-45. 3D Printers dialog box*

- Click on the **+** button from the dialog box. The **Add Printer** dialog box will be displayed; refer to Figure-46. If you have Stratasys printer or 3DSystems printer then there are direct buttons available in the dialog box to connect them with Creo Parametric.
- Select the desired option and click on the **Next** button. Follow the information given in the dialog box and create a new printer by clicking on the **Finish** button at the end in the dialog box.

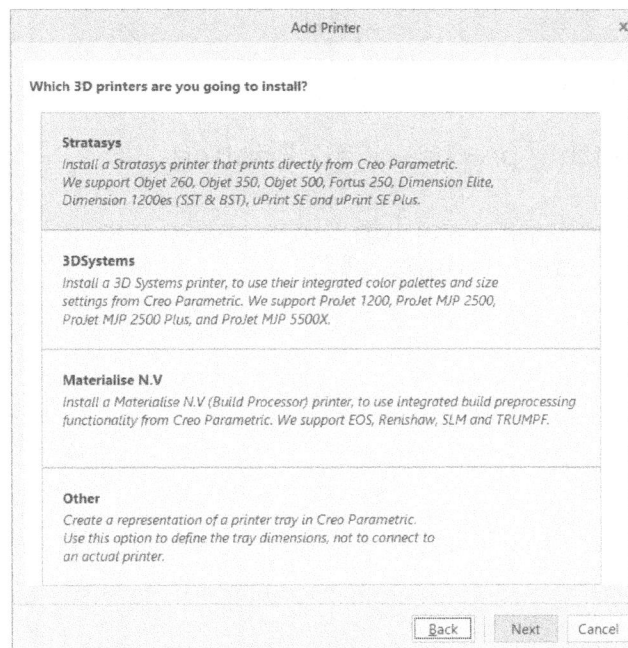

**Add Printer**

**Which 3D printers are you going to install?**

**Stratasys**
*Install a Stratasys printer that prints directly from Creo Parametric. We support Objet 260, Objet 350, Objet 500, Fortus 250, Dimension Elite, Dimension 1200es (SST & BST), uPrint SE and uPrint SE Plus.*

**3DSystems**
*Install a 3D Systems printer, to use their integrated color palettes and size settings from Creo Parametric. We support ProJet 1200, ProJet MJP 2500, ProJet MJP 2500 Plus, and ProJet MJP 5500X.*

**Materialise N.V**
*Install a Materialise N.V (Build Processor) printer, to use integrated build preprocessing functionality from Creo Parametric. We support EOS, Renishaw, SLM and TRUMPF.*

**Other**
*Create a representation of a printer tray in Creo Parametric. Use this option to define the tray dimensions, not to connect to an actual printer.*

Back   Next   Cancel

*Figure-46. Add Printer dialog box*

- After adding the printer, select it from the Printers list in the **Printer** drop-down.

## Applying Texture to the 3D Printed Part

- Select a printer from the Printers list that supports textures like **i.materialise**.
- Click on the **3D Printer Appearance** button from the **Appearance** panel in the **Tray** tab of the **Ribbon**. The **Colors** drop-down next to it will become active.

- Select the **Palette Selection** button from the **Colors** drop-down in the **Appearance** panel. The **Material Selection** box will be displayed; refer to Figure-47.

*Figure–47. Material Selection box*

- Select the desired material from the drop-down in the box and click on the **OK** button.
- Now select the desired color from the **Color** drop-down and then select the model to apply the color/texture. Click in the empty area to exit the tool.

# SELF ASSESSMENT

1. The GD&T Advisor is an application from Sigmetrix used for applying and analyzing the model for geometric dimensioning and tolerances. (T/F)

2. The **Tolerance Feature** tool is used to apply geometric tolerance and dimensioning to the whole pattern, like you can apply tolerance to the pattern of holes at single place. (T/F)

3. Explain the Vat Photopolymeriation process of 3D printing.

4. Explain the binder jetting process of 3D Printing.

5. Select the **2.5D** lattice type if you do not want to control the structure of lattice but want to define the element shape. (T/F)

# Chapter 15

# Introduction to Sheetmetal Manufacturing

## Topics Covered

The major topics covered in this chapter are:

- *Introduction*
- *Parts setting*
- *Work cell Settings*
- *Operation Setting*
- *NC Sequencing*
- *Program Output Generation*

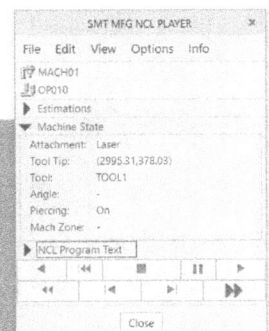

## INTRODUCTION

You have learned about creating flat pattern of various sheet metal parts in the previous chapters. You have also learned to create drawings for them. Now, you will learn about creating NC programs for automated sheet metal cutting and forming machines. In this chapter, you will learn about arranging flat patterns of various parts on a sheet so that you can maximize the usage of sheet to produce more sheet metal parts. The tools to manufacture sheet metal parts are available in the Sheet Metal Manufacturing environment. The procedure to start the manufacturing environment is given next.

*   Click on the **New** tool from the **Ribbon** or **Quick Access Toolbar**. The **New** dialog box will be displayed.
*   Select the **Manufacturing** radio button from the **Type** area and **Sheetmetal** radio button from the **Sub-type** area of the dialog box.
*   Specify the desired name in the **File name** edit box and click on the **OK** button from the dialog box. The sheet metal manufacturing environment will be displayed as shown in Figure-1.

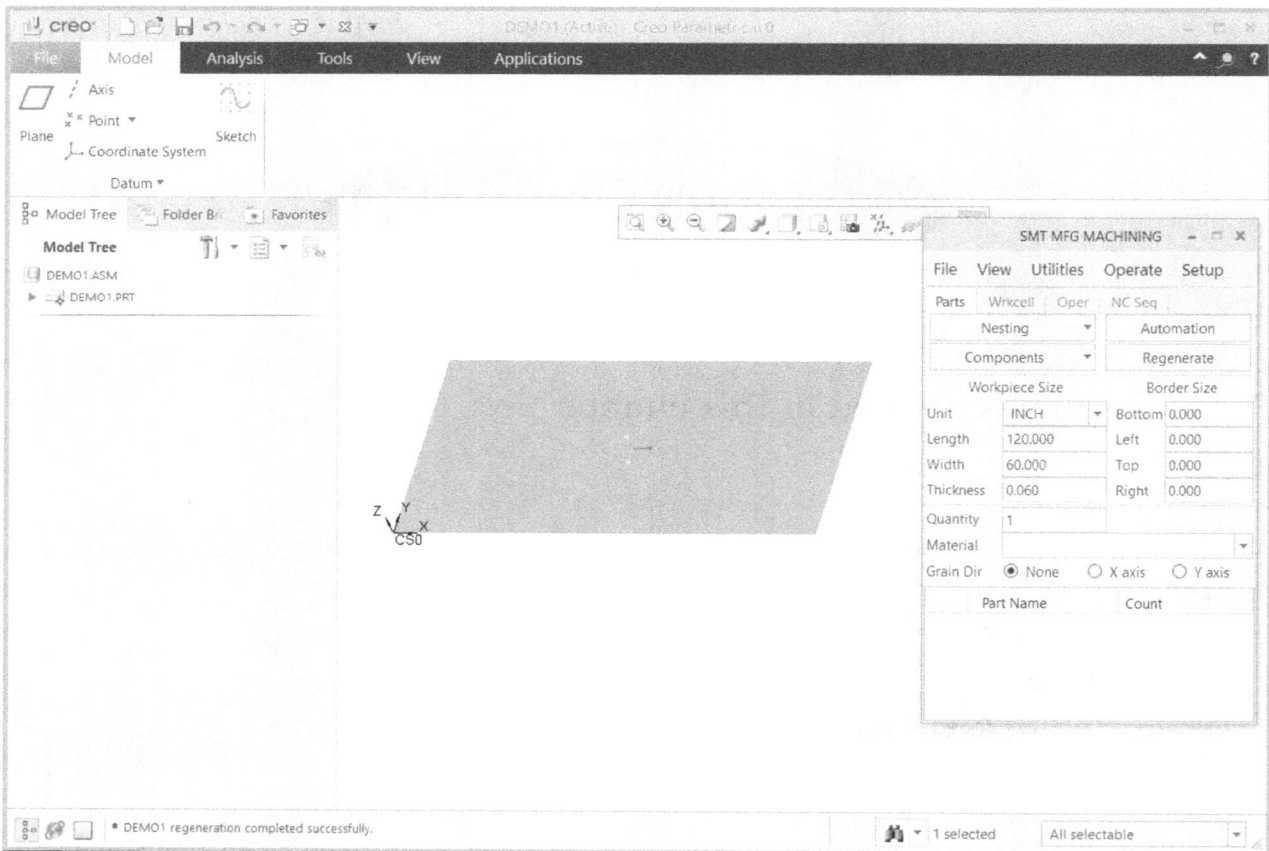

*Figure-1. Sheetmetal Manufacturing environment*

## SETTING WORKPIECE AND MACHINE

The options to set workpiece size are available in **Parts** tab of **SMT MFG MACHINING** dialog box. The procedure to set workpiece is given next.

## Setting Workpiece

* Select the desired option from the **Unit** drop-down in the **Parts** tab of the dialog box.
* Specify desired parameters in the **Length**, **Width**, and **Thickness** edit boxes to define size of metal sheet.
* Set the desired offset values in the edit boxes of **Border Size** area of the dialog box.
* Select the desired material from the **Material** drop-down and set the desired **Grain Dir** radio button.

## Setting Machine

* Select the **Wrkcell** tab from the dialog box. The options of dialog box will be displayed as shown in Figure-2.

*Figure-2. Wrkcell tab*

* Select the desired machine type from the **Type** drop-down and specify the desired name in the **Name** edit box. Select the **LASER** option from the drop-down if you want to cut flat pattern from the metal sheet. Select the **PUNCH** option from the drop-down if you want to create punch deformation in sheet metal. Select the **LASER-PUNCH** option from the drop-down if you want to perform cutting and punching both on the sheet metal. Select the **FLAME** option from the drop-down to select a flame cutting machine. Select the **FLAME-PUNCH** option from the drop-down to select flame punching machine.
* Select the **Select** option from the **Csys** drop-down at the bottom in the **Wrkcell** tab of the dialog box and select the coordinate system to be used for machine. If you want to create a new coordinate system then click on the **Create** option from the **Csys** drop-down. The **Coordinate System** dialog box will be displayed; refer to Figure-3.

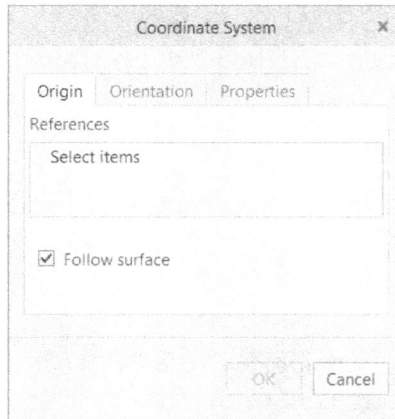

*Figure-3. Coordinate System dialog box*

- Create the coordinate system as discussed earlier and click on the **OK** button.
- Click on the **Parameters** option from the **Wrkcell** tab of the dialog box to modify parameters of machine. The **SMM PARAMETERS** dialog box will be displayed; refer to Figure-4.

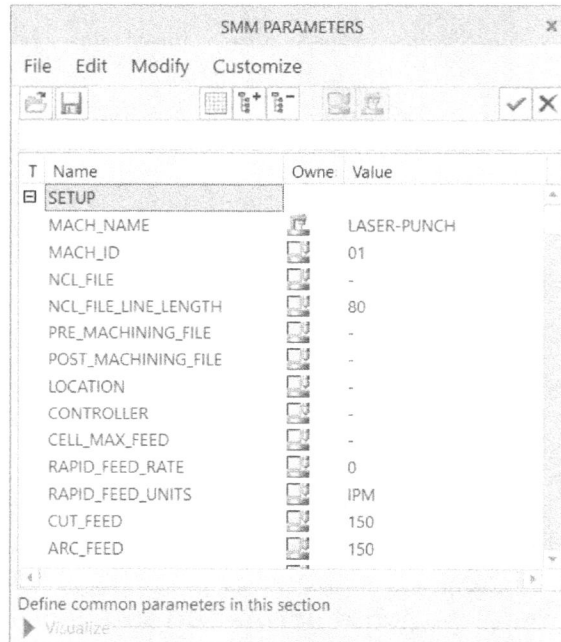

*Figure-4. SMM PARAMETERS dialog box*

- Set the desired parameters in the dialog box and click on the **OK** button.
- If you want to define a machining zone then click on the **Create** option from the **Zones** drop-down. The **Machine Zone** dialog box will be displayed with preview of machine zone; refer to Figure-5.

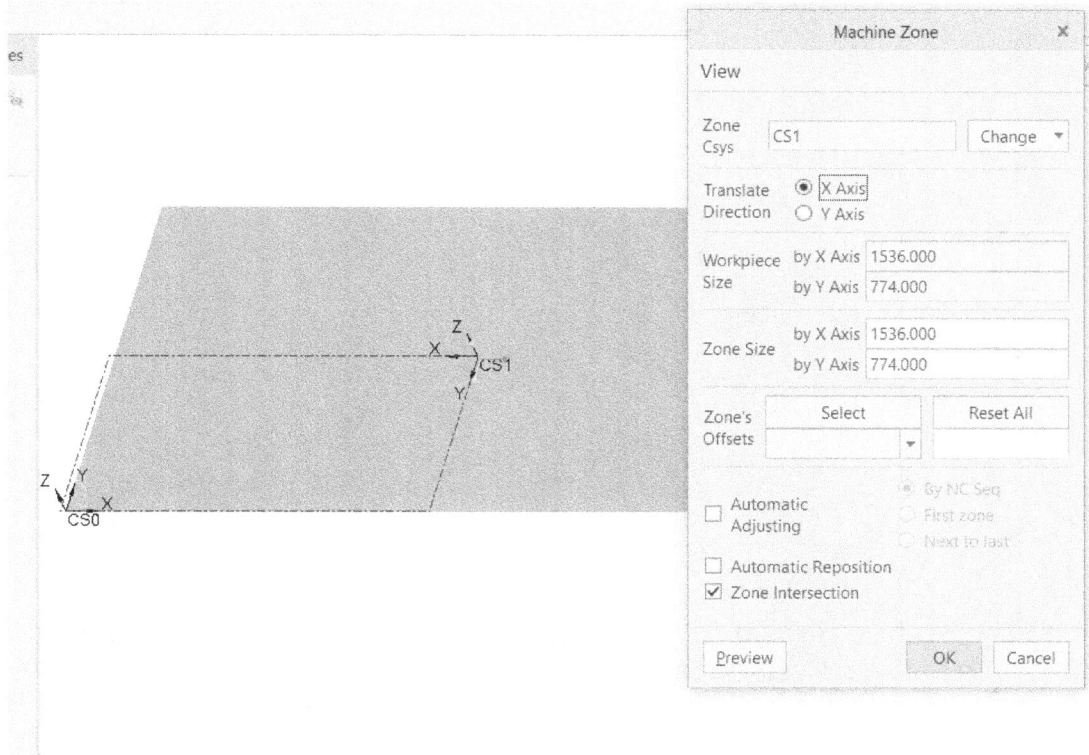

Figure-5. Machine Zone dialog box with preview of zone

- Set the desired parameters in the dialog box and click on the **OK** button.

## Tool Settings

- If you are using turret machine and want to create multiple tools for turret then click on the **Turret** button. The **TURRET MANAGER** dialog box will be displayed; refer to Figure-6.

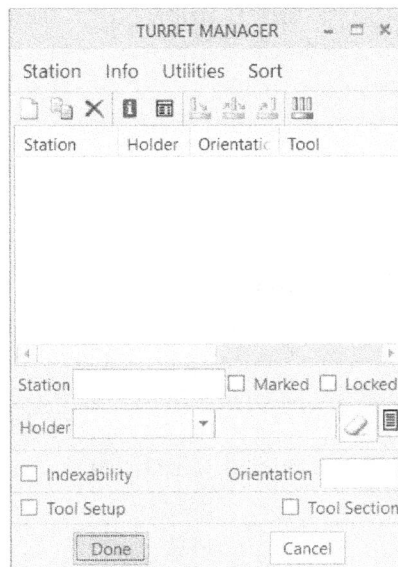

Figure-6. TURRET MANAGER dialog box

- Click on **Create New** button from the dialog box to create 1st index of tool turret. Double-click on the station to setup tool. The **TOOL SETUP** dialog box will be displayed; refer to Figure-7.

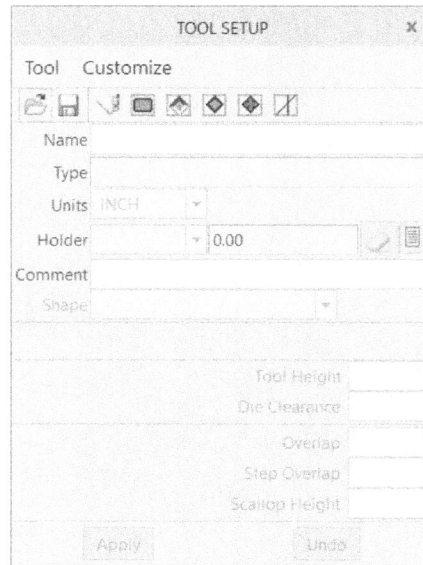

*Figure-7. TOOL SETUP dialog box*

- Click on the desired button from the dialog box. There are 6 tool options available in this dialog box viz. Contouring tool, Standard Punch tool, Form tool, UDF Punch tool, Solid Punch tool, and Shear tool; refer to Figure-8.

*Figure-8. Tools*

- The options for selected tool will be displayed in the **TOOL SETUP** dialog box. Specify the desired name of tool in the **Name** edit box. Set the other parameters as desired and click on the **Apply** button.
- Click on the **Tools/Stations Switch** tool from the toolbar in the dialog box to check tools used in machine.
- Click on the **Done** button from the dialog box to apply changes.

## Creating Operations

- Click on the **Oper** tab of the dialog box to display options related to NC operation; refer to Figure-9.
- By default OP010 is available in operations list. If you want to create a new operation like sheet cutting or punching then specify the desired name in the **Name** edit box, set the NCL file parameters, and click on the **Create** button.

*Figure-9. Oper tab of SMT MFG MACHINING dialog box*

## Part Placement

There are three options in **Parts** tab of **SMT MFG MACHINING** dialog box; Nesting, Automation, and Components. Select the **Automation** option if you want to multiple instances of same component. Select the **Components** option if you want to assemble a component with the workpiece. The procedures to perform part placement are discussed next.

## Part Placement with Nesting

Select the **Nesting** option if you want to create multiple different types of components. The procedure to use this option is given next.

- Click on the **Nesting** option from the top in the **Parts** tab of the dialog box. A drop-down will be displayed; refer to Figure-10.

*Figure-10. Nesting drop-down*

- Click on the **Create** option from the **Nesting** drop-down. The **NEST CELL Menu Manager** will be displayed.
- Click on the **Add Part** option from the **Menu Manager**. The **Open** dialog box will be displayed; refer to Figure-11.

*Figure-11. Open dialog box*

- Select the desired sheet metal part with flat pattern and click on the **Open** button. The **PART PLACE** menu will be displayed in Menu Manager. Select the **DragOrigin** check box from the menu manager to place the flat pattern by dragging and click on the **Done** button. You will be asked to select the coordinate system.
- Select the desired coordinate system and place the component at desired location.
- Click on the **Done** button from the **Menu Manager**.
- Select the desired option from the **Nesting** drop-down and modify the part placement as desired.

## Part Placement with Automation

- Click on the **Automation** button from the dialog box to create multiple instances of a sheet metal part automatically fitted on the workpiece. The **SMT MFG AUTOMATION** dialog box will be displayed; refer to Figure-12.

*Figure-12. SMT MFG AUTOMA-TION dialog box*

- Click on the **Load** option from the dialog box. The drop-down will be displayed as shown in Figure-13. There are 4 options in this drop-down; **Parts from Mfg, Cell of Parts, Part w/o NC Seqs**, and **DXF File**. Select the **Parts from Mfg** option if you want to import sheet metal part from previously created manufacturing assembly. Select the **Cell of Parts** option if you want to use previously nested parts. Select the **Part w/o NC Seqs** option from the drop-down to load sheet metal part in flat pattern. Select the **DXF File** to load flat pattern of sheet metal part created in 2D CAD software. In our case, we are selecting the **Part w/o NC Seqs** option from the drop-down. The **Open** dialog box will be displayed.

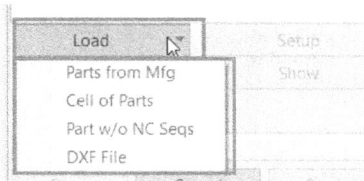

*Figure-13. Load options*

- Select the desired file from the dialog box and click on the **Open** button. A confirmation dialog box will be displayed asking you to create coordinate system for assembling the sheet metal part.
- Click in the edit box under **Order** column and specify the desired number of instances to be created for manufacturing.
- Click on the **Setup** button from the dialog box. The **Setup Part** dialog box will be displayed; refer to Figure-14.

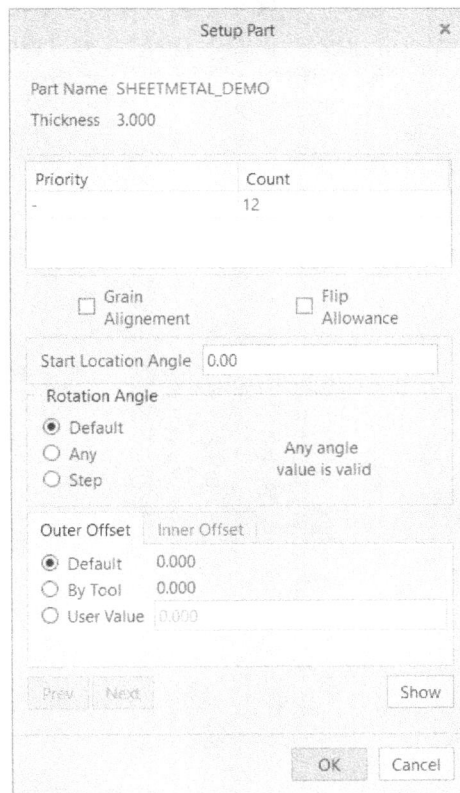

*Figure-14. Setup Part dialog box*

- Select the **User Value** radio button and specify the desired offset value. This offset will be the material left while cutting the sheet metal parts.
- Click on the **OK** button from the dialog box. The **SMT MFG AUTOMATION** dialog box will

be displayed again with number of instances specified and number of instances that could be created on specified workpiece; refer to Figure-15.

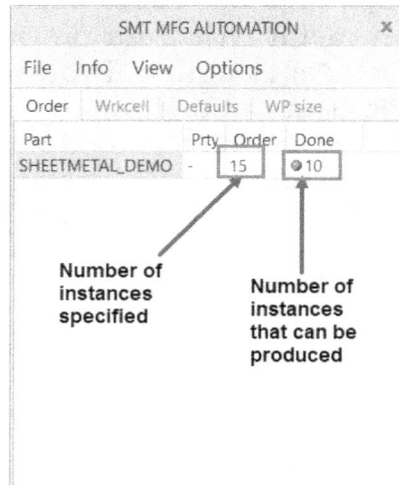

*Figure-15. Instances for sheetmetal manufacturing*

- Similarly, you can load other sheetmetal parts on same sheet if there is material left and parts can fit in there.
- Click on the **Done** button from the dialog box after specifying the parameters.

## Creating NC Sequence

The tools in the **NC Seq** tab of the dialog box are used to define toolpaths and parameters for create sheet metal part on machine; refer to Figure-16.

*Figure-16. NC Seq tab*

- Select the desired machine and operation file from the respective drop-downs at the top in the dialog box.
- The tools to create machining sequence are available below the **Operation** drop-down. These tools are:

Contouring : The contouring tool is used to cut sheet metal based on selected part boundaries.

Slitting : The slitting tool is used to split the remaining sheet metal after punching and finish cutting.

Approach Punching : The approach punching tool is used to punch the primary impression on sheet metal.

Edge Nibbling : The edge nibbling tool is used to cut contour on sheet metal by successive overlapping slits or notches.

Similarly, you can use the other tools in the dialog box to create respective toolpaths. Here, we will discuss the process of creating toolpath for contouring. You can use the same procedure for other tools.

- Click on the **Contouring** tool from the dialog box. The **TURRET MANAGER** dialog box will be displayed if you have not specified tool earlier. Create the desired contouring tool as discussed earlier.
- After selecting the tool, the options will be displayed as shown in Figure-17.

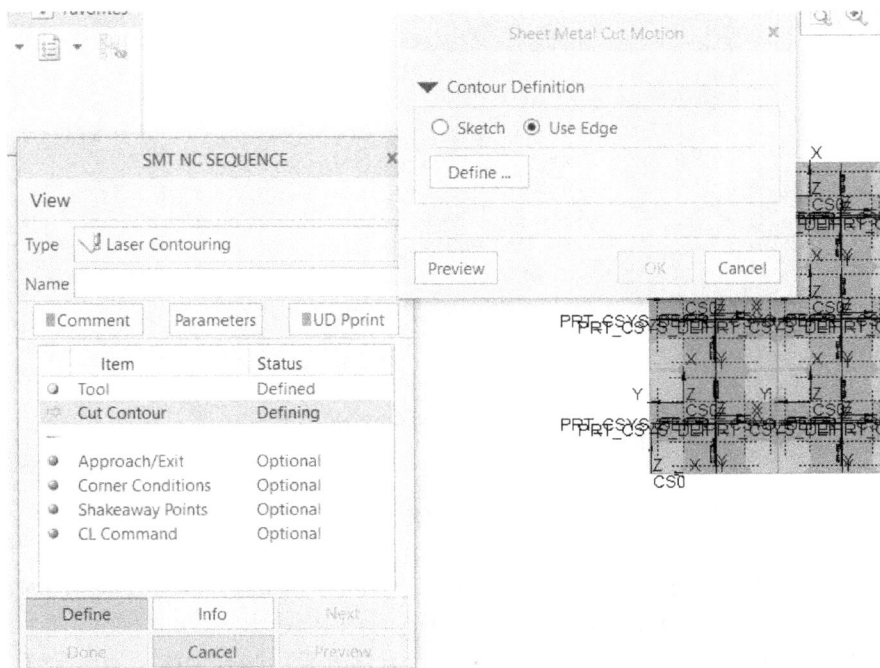

*Figure-17. Laser contouring options*

- Select the **Use Edge** radio button and click on the **Define** button from the dialog box.

- Select the edges of flat pattern part to be cut from sheet metal by contour tool; refer to Figure-18. Click on the **OK** button from the **Select Edges** dialog box. The **Sheet Metal Cut Motion** dialog box will be displayed; refer to Figure-19.

*Figure-18. Edges of flat pattern selected for contouring*

*Figure-19. Sheet Metal Cut Motion*
*dialog box*

- Click on the **Select** button in **Start Edge** area of the dialog box and select the desired edge from where you want cutting operation to begin.
- Select the **Use CutCom** check box if you want to specify the cutter compensation. Set the desired parameters after selecting the check box.
- Specify the other parameters and click on the **OK** button. The **SMT NC SEQUENCE** dialog box will be displayed.
- Double-click on the optional features in this dialog box to define them.
- Click on the **Parameters** button to define machine cutting parameters. The **SMM PARAMETERS** dialog box will be displayed as discussed earlier. Set the desired parameters and click on the **OK** button.

- Click on the **Preview** button to check toolpath. The **SMT MFG NCL PLAYER** dialog box will be displayed; refer to Figure-20.

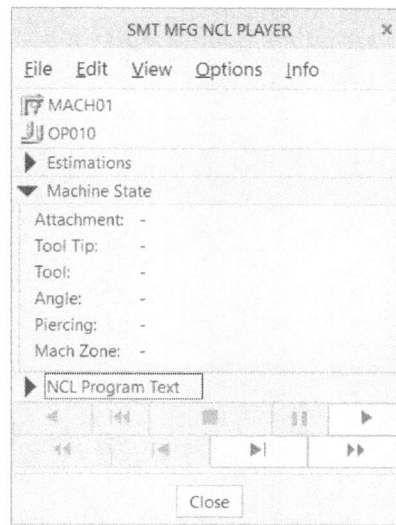

*Figure-20. SMT MFG NCL PLAY-ER dialog box*

- Click on the **Play** button in this dialog box to check the preview of toolpath. Click on the Close button after checking the preview.
- Click on the **Done** button from the **SMT NC SEQUENCE** dialog box to create the toolpath.
- Select the newly created toolpath and click on the **Create** button of **Populate** drop-down. The **SMM Populate** dialog box will be displayed; refer to Figure-21.

*Figure-21. SMM POPULATE dialog box*

- Select the toolpath once to include it. The status will be shown as included. Click on the **Preview** button and check if all instances of the part have been included.
- Click on the **OK** button from the **SMM POPULATE** dialog box to create toolpath including all instances.
- Click on the **Create** button from the **Optimize** drop-down at the bottom in the dialog box to automatically optimize toolpaths based on selected tool and machines. The **SMM Optimize** dialog box will be displayed; refer to Figure-22.
- Select the desired optimization definition like select **DEF LASER** option from the list box to optimize toolpaths for laser cutting and click on the **OK** button from the dialog box.

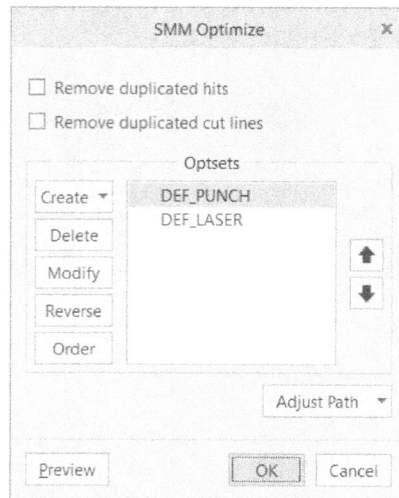

*Figure-22. SMM Optimize dialog box*

## Generating CL Output

- Select the desired NC sequence from the **SMT MFG MACHINING** dialog box and click on the **CL Output** button. The **SMT MFG NCL PLAYER** dialog box will be displayed as discussed earlier. Check the toolpath by clicking on the play button; refer to Figure-23.

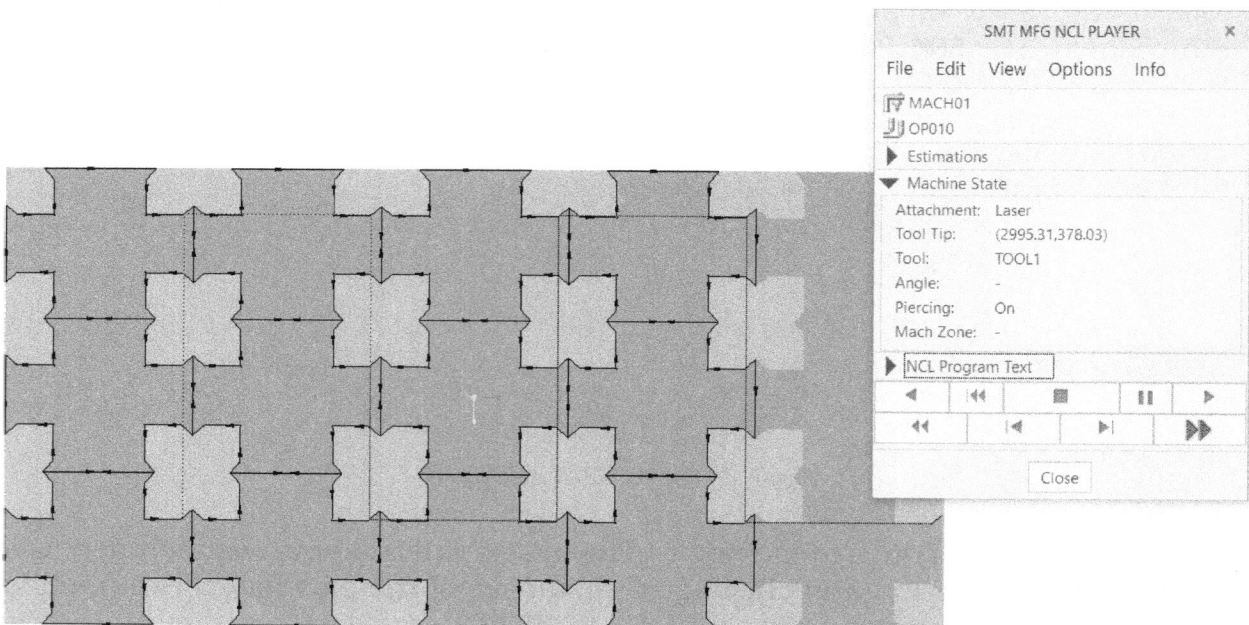

*Figure-23. Toolpath for sheetmetal contouring*

- Click on the **Save As** option from the **File** menu. The **Save a Copy** dialog box will be displayed; refer to Figure-24.

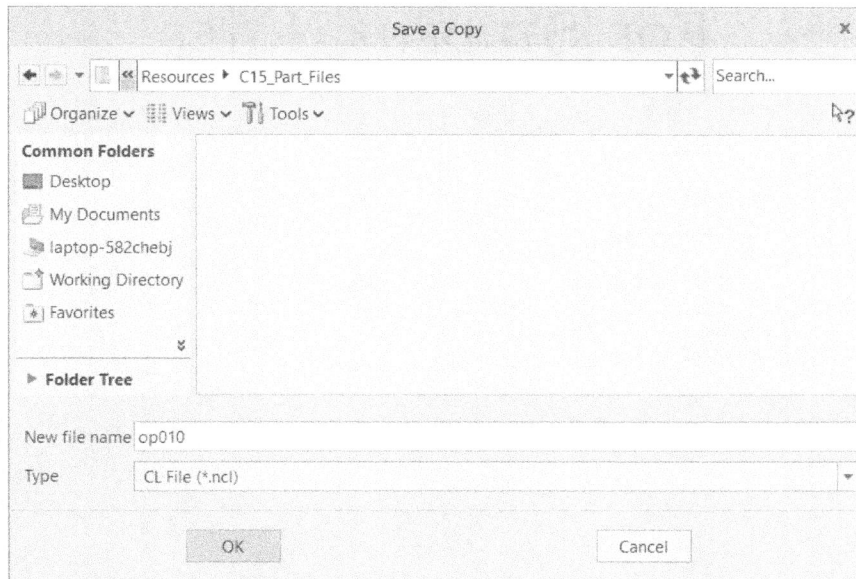

*Figure-24. Save a Copy dialog box*

- Specify the desired name and location for the file and click on the **OK** button. The output file will be created.
- Save the manufacturing file at desired location to reuse later.

FOR STUDENT NOTES

# Index

## Ethics of an Engineer

- Engineers shall hold paramount the safety, health and welfare of the public and shall strive to comply with the principles of sustainable development in the performance of their professional duties.

- Engineers shall perform services only in areas of their competence.

- Engineers shall issue public statements only in an objective and truthful manner.

- Engineers shall act in professional manners for each employer or client as faithful agents or trustees, and shall avoid conflicts of interest.

- Engineers shall build their professional reputation on the merit of their services and shall not compete unfairly with others.

- Engineers shall act in such a manner as to uphold and enhance the honor, integrity, and dignity of the engineering profession and shall act with zero-tolerance for bribery, fraud, and corruption.

- Engineers shall continue their professional development throughout their careers, and shall provide opportunities for the professional development of those engineers under their supervision.

www.ingramcontent.com/pod-product-compliance
Lightning Source LLC
Chambersburg PA
CBHW081756200326

41597CB00023B/4048